P9-APZ-965

PRAISE FOR *THE JANUS POINT*

"Julian Barbour's *The Janus Point* is simply the most important book I have read on cosmology in several years. He presents a novel approach to the central question of why time has a direction, providing a serious alternative to contemporary thinking. With a rare humanity and a perspective based on a lifetime of study of the history and philosophy of cosmology, Barbour writes a book that is both a work of literature and a masterpiece of scientific thought."

—LEE SMOLIN, author of *The Trouble with Physics*

"Julian Barbour is well known for his brilliant study of physics history, *The Discovery of Dynamics*. *The Janus Point* includes a similar history of thermodynamics, statistical mechanics, and the arrow of time. But for me the main point of the book was to show history-in-the-making. His 'shape dynamics' is a project to recast the foundations of all of cosmology, gravity, thermodynamics, and the arrow of time. The book has given me a lot to ponder. As Gauss said of Riemann's habilitation lecture, '[it] exceeded my expectations.'"

—BILL UNRUH, professor of physics at University of British Columbia

"Julian Barbour has no peer when it comes to explaining scientific ideas in a way that is accessible without being simplistic. For good measure he has a talent for using quotes from Shakespeare and other literary sources in a manner that actually helps to elucidate key points. In *The Janus Point* he tackles subject matter that is notoriously challenging even to scientists, and explains it in a way that gave me new insights and understanding even though I studied these topics in a classroom a long time ago. This is a fitting sequel to his earlier work and helps to pull together several big ideas that some of us have been watching with fascination for decades."

—NEAL STEPHENSON, author of *Snow Crash*

"By abandoning the prejudice that particles (atoms, stars . . .) are confined in a box, Julian Barbour has discovered an unexpected and remarkably simple feature of Newtonian dynamics. It is the basis of his seductive and eloquently presented explanation of the history of the universe, even time itself. Is his cosmology correct? 'Time' will tell."

—MICHAEL VICTOR BERRY,
Melville Wills Professor of Physics (Emeritus) at Bristol University

"Julian Barbour's infectious enthusiasm for the big ideas in physics is addictive. He has a complete mastery of the history of ideas yet a remarkable lightness and clarity in explaining what are profound concepts. *The Janus Point* is controversial and gripping, an extraordinary introduction to his view of the universe."

—PEDRO G. FERREIRA, author of *The Perfect Theory*

"Julian Barbour is a profound and original thinker, with the boldness to tackle some of nature's deepest problems. He is also a fine writer, and this renders his book—despite its conceptual depth—accessible to anyone who has pondered the mysteries of space and time. It's a distillation of the author's prolonged investigations, and the insights that he offers deserve wide readership."

—MARTIN REES, author of *On the Future*

THE JANUS POINT

ALSO BY JULIAN BARBOUR

The Discovery of Dynamics

The End of Time

THE JANUS POINT

A NEW THEORY OF TIME

JULIAN BARBOUR

BASIC BOOKS

New York

Cover design by Chin-Yee Lai
Cover image copyright © Nullfy via Getty Images
Cover copyright © 2020 by Hachette Book Group, Inc.

Basic Books
Hachette Book Group
1290 Avenue of the Americas, New York, NY 10104
www.basicbooks.com

Printed in the United States of America

First Edition: December 2020

Published by Basic Books, an imprint of Perseus Books, LLC, a subsidiary of Hachette Book Group, Inc. The Basic Books name and logo is a trademark of the Hachette Book Group.

Print book interior design by Jeff Williams.

Library of Congress Cataloging-in-Publication Data

Names: Barbour, Julian B., author.
Title: The Janus point: a new theory of time / Julian Barbour.
Description: First edition. | New York: Basic Books, 2020. | Includes bibliographical references and index.
Identifiers: LCCN 2020025226 | ISBN 9780465095469 (hardcover) | ISBN 9780465095490 (ebook)
Subjects: LCSH: Space and time. | Thermodynamics. | Entropy. | Second law of thermodynamics. | General relativity (Physics)
Classification: LCC QC173.59.S65 B375 2020 | DDC 530.11—dc23
LC record available at https://lccn.loc.gov/2020025226

ISBNs: 978-0-465-09546-9 (hardcover); 978-0-465-09549-0 (ebook)

LSC-C

Printing 1, 2020

In memory of my wife, Verena,
and our daughter Jessica

CONTENTS

PREFACE AND ACKNOWLEDGEMENTS

THIS BOOK'S ORIGINAL SUBTITLE was *A New Theory of Time's Arrows and the Big Bang*. That is what its substance remains, but I was happy to accept the suggestion of TJ Kelleher, my editor of the US edition, and adopt its present shorter form. I think there is warrant for it. The big bang not only gave birth to time but also stamped on it the eternal aspect of an arrow's flight. Thus the two together do amount to a new theory of time. Please note the indefinite article in the subtitle. Nothing in science is definitive. Hypotheses are proposed and tested. Science progresses when, through precise experiment and good observation, predictions are either confirmed or refuted. I believe the proposed explanation of time's arrows is as secure as is the long-established expansion of the universe, but some radical ideas about the big bang, which matured late in the writing of the book, are definitely speculative. I have nevertheless decided to include them because they do represent what seems to me, now that they have been recognised, to be an almost inevitable bringing together of everything else.

This is a very personal book in which I have tried to combine established science of the cosmos with new ideas, but I also include here and there my own reactions to existence in the universe and wonder at its nature. How is it that time has not only created the physical world of atoms and galaxies but also poets, painters, and composers? The works of Shakespeare have been a great joy in my life. You will find him quoted explicitly in a dozen or so places, but for buffs of the Bard I have also, now and then and unattributed, smuggled in from his plays and sonnets a half line or even a single word. I hope you will get a little pleasure if you spot these purloined feathers. The final chapter is not quite an epilogue because it brings in discussion of the arrow we experience most directly, that of the passage of time. I think this is intimately related to the greatest mystery of all—the gift of consciousness. Don't expect any answer to the mystery, but perhaps I can offer illumination of one of its aspects. Otherwise the chapter is an attempt to identify what it is in the mathematics of the universe that is manifested in art. It must be there since all great art has a unique structure, and structure is the very essence of mathematics.

As regards mathematics itself, inclusion of some in the book is inevitable, if only to avoid endless circumlocution. It is the concepts that count. Formulas are given for three concepts; that's all there are. Their names then make many appearances. The mere repetition may help. The renowned mathematician John von Neumann is reputed to have said, "Young man, you do not understand mathematics, you get used to it". There is a lot of truth in that. I think all my readers know perfectly well what the circumference of a circle is. With luck you will come to a similar intuition for the single most important concept in the book. I call it *complexity*. As I do here, I have used italics in the book almost exclusively to flag the first appearance of an important concept.

At the end of the book there is a list of the figures—there are twenty-six—and the page on which you will find them. Also at the end of the book, together with some additional material, I have compiled some technical notes for readers with at least some background in physics and mathematics, roughly from first-year university students to advanced researchers who want to see the evidence for statements in the body of the book and perhaps follow them up. There is also a bibliography restricted to books and technical papers that have a more or less direct relation to this one.

It includes my own *The End of Time*, published in 1999. I mention it here because some of the readers of this book will have read it and may wonder how my second venture into the genre of books for general readers differs from it. All I will say here is that the two books cover the same theme—time's arrows and their origin—but from a point of view that is somewhat different. When appropriate, I draw attention to specific differences in footnotes. There is, of course, an index.

This book could never have appeared or taken the form it does without critical input during the last eight years from my current collaborators Tim Koslowski, Flavio Mercati, and David Sloan. Some of the most important ideas come from them. It has been a great pleasure to work with them—Tim since we first met at the Perimeter Institute in Canada in 2008, Flavio since 2011, and Dave since 2015. Flavio also generated all but one of the book's figures and found online the one he did not create—the fine double-headed Janus on a Roman coin in Fig. 3. The figures are a very important part of the book and I am most grateful to Flavio for them.

To continue with acknowledgements: What we call *shape dynamics*—it's a representation of the dynamical essence of the theories of both Newton and Einstein—forms the mathematical core of this book; recent developments of it owe much to Tim, Flavio, and Dave. The notes include references to people who either directly or indirectly contributed to its earlier development, but I should mention here Bruno Bertotti (with whom I collaborated closely from the mid-1970s until 1982 and whom I was able to see in September 2018, just a month before he died), Niall Ó Murchadha, Bryan Kelleher, Brendan Foster, Edward Anderson, Sean Gryb (whom I also met at the Perimeter Institute through my friend Lee Smolin, who was his PhD supervisor there), and Henrique Gomes. Numerous visits to the Perimeter Institute, which hosted three workshops on shape dynamics, were much appreciated, as were, at the invitation of Viqar Husain, two visits (one for a workshop) at the University of New Brunswick in Fredericton. Sean Gryb also organised a valuable shape dynamics workshop at Nijmegen in the Netherlands and, with Karim Thébault, two at the University of Bristol. Two other friends of long standing with whom I have had many valuable and informative discussions are Karel Kuchař, of the University of Utah in Salt Lake City, and Christopher Isham, of Imperial College in London.

Although they are not involved in shape dynamics, I must here also thank very warmly Alain Chenciner and Alain Albouy, of the Institut de Mécanique Céleste et de Calcul des Éphémérides at the Observatory of Paris. Through the kind introduction of Richard Montgomery, of the Department of Mathematics at the University of California, Santa Cruz, I have been interacting with them on and off for twenty years. With their help I have learned much about the oldest and still in many ways the most important problem in mathematical physics; it goes right back to Isaac Newton and concerns the behaviour of point particles that interact in accordance with Newton's laws of motion and universal gravitation. It turned out that some of the most beautiful results gleaned in its study, literally over centuries, are almost tailor-made to be the foundation of much of this book. It would not be what it is without their help. To combine visits to Paris with discussion at the venerable Observatory has been a rare pleasure.

Closer to my home, Harvey Brown, working in the philosophy of science at the University of Oxford, has been a good friend and sounding board for ideas in discussions over many years; I have also had useful discussions with his colleagues Simon Saunders and Oliver Pooley. Pedro Ferreira, another professor at Oxford, in the Department of Astrophysics, has, almost inadvertently but through welcome interest in shape dynamics, played a critical role in determining the present content of my book. About five years ago he set Tim and me a challenge to see if, in the framework of Newtonian gravity represented in shape-dynamic form, we could find an alternative to what is called inflation in cosmology and is the basis of the current theory of large-scale structure formation in the universe. Pedro was also a great help in introducing us to the cosmologist Michael Joyce, who is also in Paris, though not at the Observatory. I am not going to claim we have met Pedro's challenge, but through it some entirely unexpected possibilities have come to light and figure prominently in the book. They relate to the nature of the big bang and, presented in Chapters 16 through 18, are the radical ideas mentioned above. Because they may be controversial, this is where, as the author, I must stick my colours to the mast. Hypotheses, especially if they have some plausibility, are, I think, acceptable; outright errors are quite another thing. In this book I am, especially in the ideas about the big bang and the final state of the universe, going out on something of a limb. If it breaks or I fall, any and all mistakes in the book are entirely my responsibility.

I must also warmly thank my agent, Max Brockman, of Brockman Inc in New York, and his parents, John Brockman and Katinka Matson, who handled my book *The End of Time*. I also want to very especially thank my two editors: TJ Kelleher for Basic Books in the United States (he tells me TJ is the invariable moniker by which he is known, both to family and to colleagues) and Will Hammond of Bodley Head in the UK. Having been involved with few books dealing with science for general readers, Will was glad to leave editing to TJ. Work with him has been most stimulating. He has been just what an editor should be: supportive throughout but clear about what must go or be changed.*

One major excision I will mention. I began work on this book three and a half years ago thinking I needed not only to study the history of thermodynamics, the discovery and development of which first brought to the fore the mystery of time's arrows, but also to include in the book a lot of the history, which has certainly given me a strong sense of its importance. As TJ pointed out, the first draft contained far too much; all that now remains explicitly is the minimum needed to set the scene and introduce some of the concepts and issues that feature in the heart of the book. However, there remains an arc of history that still informs the structure of the book. Moreover, several friends and colleagues were gratifyingly

---

*TJ also suggested I indicate, with extra space between paragraphs, changes of topic within a chapter greater than what you normally find between paragraphs. I liked the idea and the larger spaces are there; they mark places where you can pause, for example, for a tea or coffee break.

positive about the history and said it should see the light of day. I have therefore taken the chapters that were written but have been removed and put them as a PDF on my website, platonia.com; I hope to be able in the not too distant future to tidy them up as a book on the history of thermodynamics. You get that only in a very condensed form in this book; I hope anyone interested in the history will visit my website and perhaps download the PDF.

That brings me to one more thing: the great value of the internet and especially Wikipedia. When I next see an appeal to support Wikipedia, I do intend to make a contribution. I also recommend you check out Google Images for images of the various scientists I mention and look at Wikipedia for their biographies. It is all so easy nowadays; I think you will find it well worth the minor effort of a click.

Mention of modern digital technology brings me to thank Melissa Veronesi of Basic Books for her help with final checking of the text as it came to me from the copy editor, Sue Warga, who has done a great job. It has been an eye-opener to see in how many ways text can be made to flow better. I am also very grateful to my son, Boris, for help with digital issues and preparation of the text.

Boris is not the only member of my family I want to thank. My wife, Verena, succumbed to Alzheimer's just after I submitted the proposal for this book to my agent in June 2016; she had remained wonderfully cheerful to the end. But in the midst of the writing of this book, eighteen months after her mother's death, I lost my daughter Jessica. Her two sisters, Naomi and Dorcas, together with Boris, have been a great comfort, as have eight grandchildren, including Jessica's two daughters. You will see that the book is dedicated to the memory of Verena and Jessica. Both very greatly enriched my life.

CHAPTER 1

TIME AND ITS ARROWS

The Universe is made of stories, not of atoms.

—MURIEL RUKEYSER

TIME FLOWS FORWARD. EVERYONE HAS THE FEELING. IT IS MORE THAN MERE feeling; it has a real counterpart in observable phenomena. Processes near and far in space and time all unfold in the same direction. All animals, us humans included, get older in the same direction. We never meet anyone getting younger. Astronomers have observed millions of stars and understand very well how they age—all in the same direction as us. On seashores around the world, waves build and break—they never 'unbreak'.

We remember the past, not the future. There are arrows of time. A film run backwards confounds cause and effect: instead of divers making a splash on entering the water, they emerge from it while the splash disappears. Myriad arrows permeate our existence. They are the stuff of birth, life, and death. In their totality, they define for us the direction of time.

Three arrows are particularly important because they can be treated with a good degree of mathematical precision.

The first, with which we will be much concerned, is the common process of equilibration. To see an example and its end result, put a tumbler of water on a table and disturb its surface by stirring the water vigorously with your finger. Remove your finger. Very soon the disturbance subsides and the surface becomes flat. You have witnessed an irreversible process. You can watch the surface for hours on end and it will never become disturbed spontaneously. The observationally unchanging state reached through equilibration corresponds to equilibrium. This is an example you can see. More important for the subject of this book is one you can feel—the equalisation of temperature. Go from a warm room to a cold room and you immediately feel the difference because your body is losing heat to the air around it. Many similar examples could be given: hot coffee in a mug, if not drunk, cools to the ambient temperature of the room.

The next arrow relates to what are called retarded waves. Don't worry about the name. You see a beautiful example whenever you throw a stone into a still pond and circular waves spread out from the point of entry. The waves are said to be retarded because they are observed after the impact of the stone—the effect follows the cause. You never see waves that mysteriously start in unison at the bank of the pond, converge on its centre, and eject a stone that then lands in your hand—although the laws of hydrodynamics and mechanics are perfectly compatible with the possibility. It is not only water waves that invariably exhibit retardation; so do the radio and TV signals that reach your home from transmitters.

The third example is in quantum mechanics and is the notorious problem of Schrödinger's cat. In accordance with the quantum formalism, a certain wave function can describe a cat that is simultaneously both dead and alive. Only when an observation is made to establish the cat's state is there collapse of the wave function and just one of the two possibilities is realised.

Arthur Eddington, the British astrophysicist who in 1919 made Einstein into a world celebrity overnight with a famous telegramme to the London *Times* that confirmed a prediction of the general theory of relativity, coined the expression 'arrow of time' in the 1920s. Eddington had in mind mainly the arrow associated with equilibration, but people often use it as a portmanteau expression to cover all the arrows, as I will.

BECAUSE THE ARROWS of time are so ubiquitous it's easy to take them for granted, but since the early 1850s theoreticians have seen in them a major problem. It is easily stated: apart from a tiny temporal asymmetry in one single law that, just after the big bang, was one of the factors that prevented complete mutual annihilation of matter and antimatter but cannot have played any significant role in creating the currently observed huge asymmetry between past and future, all the remaining laws of nature make no distinction between past and future. They work equally well in both time directions. Take a very simple law, the one that governs billiard balls. Unlike divers plunging into water, a film that spans their impact does look the same forwards and backwards. Physicists say laws like that are time-reversal symmetric. By this, they do not mean that time is reversed but that if at some stage all relevant velocities are precisely reversed, the balls will retrace, at the same speeds, the paths previously taken. More complicated is the case of charged particles in a magnetic field; the direction of the field must also be reversed along with the particle velocities if the retracing is to occur. There is an even more subtle case related to electric charges, mirror reflection, and the single fortuitous rule-breaking exception which allowed a minuscule amount of matter to survive after the big bang and we who are made of it to come into existence billions of years later. In all cases, the problem is the same: if the individual laws of nature that count do not distinguish a direction of time, how does it come about that innumerable phenomena do?

This question throws up such basic issues that we cannot hope to answer it unless we identify secure foundations for our theorising. One such foundation is that all meaningful statements in science are about relationships. This applies to the very notion of the direction of time. We recognise one because all around us we have those multitudinous unidirectional arrows. Without their constant presence, a single diver emerging backwards out of turbulent water, leaving it smooth and flat, would not appear to violate the normal course of nature. Also important are the nature and precision of observation. Wearing night-vision spectacles sensitive to infrared light, we would not see the collision of billiard balls as the same backwards and forwards—we would see hot spots appear on both balls after the collision, while the film run backwards would show those spots disappearing. In cricket, infrared imaging cameras are used when umpires are in doubt whether the ball has nicked the bat or pad and, through friction, raised the temperature at a localised spot.

This sensitivity to the means of observation raises a critical question: are the laws of nature truly time-reversal symmetric? The answer almost universally given is that yes, they are at the fundamental level of elementary particles—there is no microscopic arrow of time, only one at the macroscopic level. This was the conclusion that scientists reached in the second half of the nineteenth century. It came about in the first place through one truly remarkable study. Towards the end of the eighteenth century, people started to investigate seriously the properties of heat. A large part of the stimulus for this was to understand the workings of steam engines and how to make them as efficient as possible. In 1824, the young French engineer Sadi Carnot published a book, remarkable for its profundity and brevity, on this problem. Initially it passed unnoticed, but in 1849 a paper by William Thomson (later Lord Kelvin) brought Carnot's work to the notice of Rudolf Clausius, with dramatic effect: together with Thomson, Clausius played a key role in the creation of an entirely new science, for which Thomson coined the name thermodynamics.

Its first law states that energy can be neither created nor destroyed. Within physics, this formalises the long-held belief reflected in Lear's warning to Cordelia: "Nothing will come of nothing". The second law of thermodynamics introduces the fundamental concept of entropy, a great discovery by Clausius, who introduced the name. Entropy is, to say the least, a subtle notion that, since

Clausius's discovery, has been formulated in many different ways. Indeed, a search through the technical literature throws up about twenty different definitions of the second law. Some of them include entropy; others do not. Despite that, entropy is one of those scientific concepts that have entered common parlance along with energy, the big bang, the expansion of the universe, black holes, evolution, DNA, and a few more. In a non-technical simplification that can mislead, entropy is widely described as a measure of disorder. The most common formulation of the second law states that, under controlled conditions that allow proper definition, entropy cannot decrease and generally will increase. In particular, entropy always increases in any process of equilibration in a confined space. Among the various arrows of time, entropy's growth is the one that the majority of scientists take to be the most fundamental. The second law, often referred to without the addition 'of thermodynamics', has acquired an almost inviolable aura. In quite large part this is due to Clausius, who knew how to make sure that, deservedly, his results lodged securely in the mind. The paper of 1865 in which he coined 'entropy' ends with the words "Die Entropie der Welt strebt einem Maximum zu" (the entropy of the universe tends to a maximum). This statement had the impact that Clausius intended and was taken to mean that the universe will eventually expire in the heat death of thermal equilibrium. In human terms, there's a near anticipation of this in the final quatrain of Shakespeare's Sonnet 73:

> In me thou see'st the glowing of such fire,
> That on the ashes of his youth doth lie,
> As the death-bed whereon it must expire,
> Consum'd with that which it was nourish'd by.

Not surprisingly, the second law is grist for the mill of all pessimists. For many scientists, the growth of entropy, and with it disorder, is the ineluctable arrow that puts the direction into time. The mystery then is how the arrow gets into things so profoundly if the laws give no indication that it should.

THE DIFFICULTY RESIDES in the structure of the laws. By themselves they do not tell you what will happen in any given situation. You have to specify an initial condition. In the billiard example, you need to say where the two balls are at an initial time and the velocities they have at that moment. The law will then tell you what subsequently happens. It will give you a solution. The situation is this: *Law + Initial Condition → Solution.* In the simplest billiard example, the solution is time-reversal symmetric, just like the law that creates it. But the examples of processes with which I began this book most definitely are not. As I indicated then, scientists have not yet been able to resolve the mismatch between the symmetric laws and the asymmetric solutions.

What they have been able to do is at best give a partial solution. To illustrate this, one needs a game with more balls than billiards: snooker. The game begins with the white cue ball striking the triangle of fifteen reds. Successive impacts send the balls flying in all directions. No individual two-ball impact distinguishes a direction of time. But what then happens does. That's where the law which at the elementary two-ball (microscopic) level describes a phenomenon without an arrow is able to create a many-ball effect with a macroscopic arrow. Theoreticians are universally in agreement that many dynamical 'agents' (the balls in snooker, elementary particles in fundamental physics) must be present if a macroscopic arrow is to emerge from time-symmetric microscopic laws.

However, this is at best a necessary condition; by itself it is not sufficient. To understand this, make a film of the initial impact in a game of snooker, its explosive effect, and a few bounces off the table walls. Run it backwards. Each individual impact again satisfies the reversible billiard law. But, in a seeming miracle, reds with apparently random speeds and directions of motion conspire in the midst of their chaos to come to rest in a perfect triangle and eject the white.

Of course, there is no miracle. The game begins with a special initial condition. That singles out a special solution. Only the tiniest fraction of all possible solutions is like that. But the example does at least show that reversible laws are compatible with solutions like 'unbreaking' waves, provided that sufficiently many objects are involved. There is no absolutely irreconcilable conflict with the laws of nature even if one has to invoke something close to divine intervention. De facto, this is what physicists are forced to do. They have to assume that at some time in the distant past, most probably at the big bang, the universe was in a special state of extraordinarily high order, the ultimate origin of all time's arrows.

The philosopher of science David Albert has dubbed this assumption the 'past hypothesis'. It was first proposed by the great Austrian physicist Ludwig Boltzmann in the 1890s as one possible way to explain what is now called the arrow of time. But it's a stopgap he did not favour because it does not respect temporal symmetry. Science aims to explain phenomena by laws, not by inexplicable initial conditions arbitrarily imposed. By that criterion, the hypothesis fails, as I think Albert himself would not deny. But this leaves us in a very uncomfortable place: the most profound aspects of existence are attributed not to law but to a special condition 'put in by hand'. It's not a resolution; it's an admission of defeat. However, I would not be writing this book if that were all I had to say. Plenty of other writers have already done a good job of describing the problem; their books are included in the bibliography and there are brief comments about them in the notes to this chapter.

INSTEAD, I'M GOING to suggest that the problem could have a genuine—and surprisingly simple—resolution. My collaborators and I stumbled on it a few years ago while working on a different problem. In this chapter I will give you an outline of our proposal and then fill out the necessary details in the chapters that follow. If the proposals that I present are on the right lines, I think that, taken together, they do amount to a new theory of time itself and not just its arrows. While the arrows by themselves represent a major aspect of time and a huge part of our deepest and most intimate experience, a proper understanding of their nature is impossible without a radical transformation of our notion of time. Accordingly, this book has two parts. The first sets the scene with a brief history of thermodynamics, formulates its principles, and explains why they fail to solve the problem of time's arrows. The second presents the proposed solution; it takes up the bulk of the book.

We need the history because thermodynamics grew out of a specific problem which arose at a particular point in time: the industrial revolution and Sadi Carnot's search for the most efficient way to operate steam engines. In that slim booklet published in 1824 he laid down principles of remarkable robustness—they have stood now for almost two hundred years—but they all apply in a 'box scenario', that is, to steam, gas, or any fluid in an impermeable and insulating container. Despite this critical condition, it is widely assumed that the laws of thermodynamics and the notion of entropy can be carried over more or less unchanged to cosmology. But the universe is not in a box; it is expanding, seemingly without impediment.

The steam engine cylinder was the fruitful womb of thermodynamics, not of time's arrows. The mother and birth pangs of the latter must be sought in the universe, in which principles of far greater power than those of steam engines hold sway. Carnot sought the most efficient use of coal, not its ultimate origin. Of course, he lived too early for that; we are better placed to take the principles of thermodynamics out of their box. For we know something which neither Carnot, Boltzmann, nor any other scientist in the nineteenth century had any inkling of: that the universe is expanding. That changes everything. The box-scenario principles must, without loss of power, be turned almost upside down to describe a universe that can expand freely. Only then can the mystery of time's arrows be solved.

In the next two chapters I will describe the history of thermodynamics and explain, as simply as I can, the original form of its principles so that in the new context their essential core can, despite the radical inversion, be seen to be intact and to suggest possibilities of which Boltzmann never could have dreamed.

The Copernican revolution is the prime example of the knock-on consequences a change of perspective can unleash. The great comet of 1577 led to one of the first, and it shows what can happen if a conceptual box is swept away. Around 1600 Johannes Kepler challenged the notion that crystal spheres carried the planets, arguing that Tycho Brahe's observations proved the comet to be so far away that if the crystal spheres existed, it would have had to crash into them. Hence the spheres could not exist. That prompted Kepler to comment with far-reaching effect: "From now on", he said, "the planets must find their way through the void like the birds through the air. We must philosophise about these things differently".

He did, and it led him to his laws of planetary motion, without which Isaac Newton never could have found his law of universal gravitation. That in turn set the scientific revolution on its as yet unstoppable course. The analogy between a physical (or conceptual) box and crystal spheres is not perfect, but it will do. Both impose constraints. Freed from the spheres, Kepler's planets could go their own way; freed from a box, atoms can fly off in all directions.

Copernicus's and Kepler's examples also show how even though a conceptual framework may be overthrown, its technical achievements may live on in a new guise. Copernicus did not introduce any new mathematical techniques when he set the earth in motion through the heavens. He took all the techniques he used from Ptolemy's *Almagest*, written about fourteen hundred years earlier. Kepler too used those very same methods time and again. His discoveries are unthinkable without them.

Thus, we too can 'think outside the box' while still using what was learned within it, suitably modified. Luckily, this can be done with a model barely more complicated than the simple ideal-gas model that was used, with great success, to give an atomistic explanation of gas thermodynamics and led to the discovery of entropy. Little more needs to be done than adapt it to a system like the universe, which current observations suggest will expand forever. This leads to the two major changes to current thinking that I propose.

THE FIRST, AN idea that came almost by happenstance, is that the big bang ceases to be an explosive birth of the universe and with it time. Instead it is a special point on the timeline of the universe, on either side of which the universe's size increases. I call this special location the Janus point—hence the title of the book. Everything related to this idea is based on long-known physics but, critically, interpreted in terms of ratios. The reason this helps is that general relativity, as usually interpreted, is said to break down precisely at the instant at which the universe comes into existence at the big bang. As we look back in time to that moment, Einstein's theory predicts that the size of the universe becomes zero and that matter densities and other quantities become infinite. Physicists take such infinities as a sign that something must be wrong with the theory. It is said to predict its own demise.

However, at any instant, the universe has both a size and a shape, and the latter is defined in terms of ratios. The difference is important. A triangle gives a simple illustration of the distinction between size and shape. From the lengths of the three sides, mathematicians form ratios that determine the internal angles of the triangle, which fix its shape. If you hold up a cardboard triangle in front of you with its surface at right angles to your line of sight and move the triangle toward and away from your eyes, I think you will agree that the shape is what truly characterises the triangle's intrinsic nature.* The size is relative, and it appears to change only because your external viewpoint does. Ants on the triangle could only observe how its size changes relative to them and determine how the 'universe' they form with it changes. Similarly in cosmology, only the variables which describe shape are observable and hence physical. Astronomers do not see the universe expanding; they see it changing its shape and from that deduce its expansion.

I need to say here that Janus points of two different kinds can exist: either the size of the universe becomes exactly zero or it merely passes through a minimal value. The latter situation, in

---

* Readers familiar with my earlier book *The End of Time* may recall my brief introduction of the notion of the shape space of the universe, which, with collaborators, I was just beginning to study. The ideas presented in this book, which have taken twenty years to mature and continue to give surprises, owe much to the notion of shape space. I am now persuaded that it, rather than Platonia (which is shape space with the notion of size added), should be regarded as the fundamental arena of the universe.

which the shape variables can be continued smoothly and uniquely through the Janus point to 'another universe' on the other side, is already fully adequate to explain the arrow of time, and it gives rise to many striking and interesting effects. Most of the chapters devoted to the Janus-point idea describe those effects. The second possibility, that the size goes to zero exactly, has consequences that potentially are considerably more far-reaching and suggest a different kind of Janus point, a veritable big bang with multiple universes on either side of it. Moreover, they lead to a state at the Janus point that looks suspiciously like a past hypothesis of the kind I just said was only a stopgap. There is, however, a difference. First, it is, so far as I know, the first explicit example given of such a state; second, its very form and the manner in which it arises suggest that it is not 'put in by hand' but arises as a necessary consequence of a *law of the universe* that is more fundamental than the individual laws of nature and indeed from which they all emerge as approximate descriptions of locally observed phenomena. Unfortunately, the study of the zero-size case involves mathematical issues that are far from trivial, and I devote to it less space, in part at least because I cannot go beyond conjectures. However, those conjectures are of such interest that I do believe they deserve inclusion.

The second change to current thinking, prompted by the first, is the introduction of an entropy-like quantity—called entaxy, to distinguish it from conventional entropy defined for boxed systems—suitable for an 'unboxed universe'. In contrast to ordinary entropy, it does not increase but decreases. This does not conflict with the second law of thermodynamics, which—as proved by Clausius—applies only to confined systems within the universe. The decrease of this new quantity as the universe evolves reflects above all the growth not of disorder but of *complexity*, manifested as the birth and growth of structures and with them the formation of previously nonexistent subsystems that become effectively self-confined. It is within such subsystems that conventional entropy increases. In fact, complexity, defined by a formal mathematical expression, is a more important notion than entaxy; it reflects what we see in the universe and makes it possible to define entaxy.

Because of the way the structure in the universe, represented as complexity, evolves on the two sides of the Janus point, the notion of time is transformed. It no longer has one direction, from past to future, but instead has two: from a common past at the Janus point to two futures in the two directions away from it. Two *back-to-back* half-solutions, each a 'half-Janus', describe the universe. A divinity able to survey the whole of the universe's timeline would see the growth of its complexity, manifested in the formation of localised structures—the subsystems just mentioned—in the same two directions.

If growth of structure defines the direction of time, the divinity cannot fail to see it flip at the Janus point. From that special location, she could, like Janus, look in the two opposite directions of time at once. She would see the same kinds of thing happening both ways, above all structure being created out of near uniformity. The various arrows of time that I described at the start of this chapter would all be there, in both cases as if in flight from Janus. She would see that, in its totality, the universe respects the symmetry of the law that governs it; there is the same kind of behaviour on both sides of the threshold on which Janus traditionally stands. But should she, like Zeus's daughter Athena, come down from her Olympian vantage point to mix among mortals,* who can only exist on one side or other of the Janus point, she would find these ordinary beings perplexed. They cannot look through the Janus point and see what is on its other side; restricted in their vision, all they find around them are time-reversal-symmetric local laws of nature that, by some alchemy they call the past hypothesis, give rise to universally unidirectional arrows of time. But there is no alchemy. The law of the universe dictates the existence of the Janus point in all its solutions; there is symmetry overall, not in the particular.

---

*A recent read of Emily Wilson's splendid translation of Homer's *Odyssey* prompted me to make Athena the divinity; unlike her father, who regularly escaped the wrathful eye of his consort, Hera, in Olympus to seduce beautiful women, Athena invariably came to the aid of her favoured Odysseus when he was in dire need; the relationship was entirely chaste.

BOTH WAYS IN which a Janus point can be realised, with or without the size of the universe becoming zero, require a radically new way of thinking about the nature of time and the universe. I introduced a little earlier a major theme in the book: that instead of regarding locally determined laws of nature as fundamental, we should instead recognise them as manifestations of a single, overarching law of the universe. There is justification for this. Ernst Mach, the great nineteenth-century physicist whose discovery of shock waves is honoured by the eponymous numbers measuring them, argued in a famous book on mechanics, published in 1883, that without historical investigation of science, "the principles treasured up in it become a system of half-understood prescripts, or worse, a system of prejudices. Historical investigation", he said, "not only promotes the understanding of that which now is, but also brings new possibilities before us, by showing that which exists to be in great measure conventional and accidental. From the higher point of view at which different paths of thought converge we may look about us with freer vision and discover routes before unknown". Mach's book, with its critique of Newton's notions of absolute space and time, was a major stimulus behind Einstein's creation of the general theory of relativity between 1907 and 1915 as well as Heisenberg's formulation of quantum mechanics in 1925.

So I take Mach's admonition as encouragement for my investigation of the history of thermodynamics and the arrows of time. And the history matters. The purely practical and contingent circumstance of the birth of thermodynamics as a theory to explain the workings of steam engines has had a lasting effect: almost without exception, some means of confinement has, ever since Carnot's pioneering study, played a critical role in the study and discussion of entropy and its associated arrow of time. In experimental work some real physical container, whose size can often be changed within finite limits (as when a piston is pushed into a cylinder), has been employed; in theoretical studies a box has been modelled as an essential part of the conceptual framework. However, results obtained within an enclosure are still widely and erroneously assumed to be valid in situations without an enclosure, even by scientists as distinguished as the late Richard Feynman. The only possible explanation for this is a lack of awareness of the critical condition—confinement and insulation—that must hold. If Feynman got it wrong, we might need to reconsider some things along the lines Mach advocated.

This does not mean blanket rejection of confinement. Far from it—an enclosure, both physical and conceptual, made possible some of the most profound discoveries in science, including unambiguous evidence for the existence and size of atoms and, in 1900, the discovery of quantum effects. The wonder and importance of that can hardly be overemphasised. But I will explain why, seventy years or so after Carnot published his book, the box scenario thwarted Boltzmann's heroic attempts to give a completely convincing mechanical explanation of entropy's increase. And I will argue that now, another 120 years on, a box mentality may still be blocking the way, through the Janus-point idea, to a true explanation not only of entropy growth but of all of time's arrows.

If correct, this idea will resolve what appears to be a flagrant contradiction: the current consensus is that, for some as yet inexplicable reason, the universe began with very low entropy and a correspondingly high order that has since been undergoing inevitable progressive degradation through the inescapable effect of the second law of thermodynamics, yet all around us in the universe we see a wonderful growth of structure and order. One main aim of this book is to cast doubt on the widely accepted account of what has been happening throughout cosmic history, or at least on the notion of progressive degradation. I will use the example of records—for example, documents, fossils, and stars (considered as 'fossils' of stellar evolution)—which have remarkable mutual consistency, and give evidence for quite the opposite of what is claimed: that the history of the universe is not one of increasing disorder but rather of the growth of structure.

TO MAKE THIS case, let me begin with a personal example. In 1963 I was very lucky to buy at auction a beautiful house in the village where I live, South Newington. It is called College Farm because, prior to my purchase, it belonged to New College in Oxford for about a century. It had been built just at the time when the style of yeoman farmhouses had evolved to near perfection. That's a very local example of structure development that I have now enjoyed for decades. High on the house front is a date stone with the letters *EB* above a finely carved *1659*, with the letters *DB* below. In the charming *Book of Newton*, written by an amateur historian in the 1920s, I found the claim that *EB* stands for Edward Box, the farmer said to be the first owner of the house, and *DB* for his wife, Dorcas Box.

I thought it would be interesting to check it out, and so I headed to Oxford and its famous Bodleian Library (often simply called Bodley, after Sir Thomas Bodley, who rescued it from neglect between 1598 and 1602). The library has an immense store of books and ancient documents, some of them on open shelves in the fine 1488 reading room named after the original founder, Duke Humfrey. He was the younger brother of Henry V, having fought alongside him at Agincourt on St Crispin's Day in 1415 as one of the happy "band of brothers". Consulting the catalogue in the reading room, I found the library held the will and inventory of one Edward Box, who had lived in South Newton. That had to be him; place names can change. I filled out an order slip.

After about fifteen minutes a wooden lift, hauled by a rope pulley, came rumbling up from the bowels of Bodley.* It bore the requested bundle of documents. They clinched the authenticity of the village history: the inventory listed all Edward's chattels, room by room. The rooms matched, one for one, the existing rooms and cellar of the house, including the cheese garret and the north garret, in which, at this very moment, I sit and type these words.

This is just one example of many that convince us the past was real. Records we can trust are legion and distributed all over the world, in the solar system, in the galaxy, and throughout the whole

---

*The lift, running up the centre of the north stairwell in Bodley, was removed some years ago, but the holes of screws that held the lift walls can still be seen. Oxford being what it is, they may well be there centuries hence for any curious reader who happens to read these words and be in Oxford.

of the universe as far as we can observe it. They are present in an effectively permanent form and are mutually consistent in a manner that goes way, way beyond my experience. The example of my house's past was already convincing because different pieces of evidence—the date stone, the village history, the details in the documents, and the actual rooms in the house—cohered so well. Consistency like that is the basis of Hippolyta's counterargument to Theseus's dismissal of the Athenian lovers' narrative of their midsummer adventures as fantasies of seething brains:

> But all the story of the night told over,
> And all their minds transfigured so together,
> More witnesseth than fancy's images
> And grows to something of great constancy.

But what we find around us in the world and the universe at large goes far beyond what Shakespeare gives us through the magic of his wonderfully constructed play.

An excellent example is found in the work of geologists. In the late eighteenth century, most Europeans and Americans still believed the world had been created as described in Genesis. Many took seriously the 'calculation' by James Ussher, the Protestant archbishop of Armagh and primate of all Ireland between 1625 and 1656, that creation took place at "the entrance of the night preceding the 23rd day of October, the year before Christ 4004". However, geologists were beginning to interpret evidence in the form of rocks, fossils, the not-yet-understood effects of glaciation, and the rate at which mountains are eroded by weather into sand and carried grain by grain down rivers to the oceans. Fossils were particularly striking. They could be found at different depths in sedimentary rock strata that are identical in terms of mineral content (because laid down under almost the same physical conditions), with the fossils changing little by little according to depth but clearly related. How, the geologists asked, could the earth have come to its striking form?

They could get some estimate of the age of mountains from their calculations of the effect of weathering, while the fossil record strongly suggested evolution in time.* This was how the geologists came to 'deep time'. A typical reaction to the geological evidence was that of Jean-Baptiste Lamarck, the well-known precursor of Darwin as a theorist of evolution. He asserted that "the earth's surface is its own historian".

The consistency of the record deduced from fossils and the levels of sedimentary rock strata must have some explanation. It was sought in the hypothesis that the earth had evolved subject to laws, at that time only imperfectly sensed, for an immense length of time. In fact, in the second half of the nineteenth century this led to a famous controversy about the age of the earth between the geologists (and supporters of Charles Darwin's theory of evolution, published in 1859) and the leading British physicist of the day: William Thomson, who argued for an age of only about 20 million years. In fact, despite his preeminence, Thomson lost the argument comprehensively, albeit posthumously, through development of dating based on radioactive decay, which gives the earth an age of more than 4.5 billion years—250 times greater than Thomson's estimate.

At the terrestrial level, fossils provide good examples of the kind of evidence we find throughout the universe. First, when they are considered as records of the past, they are found upon careful examination to be mutually consistent to an extraordinary degree—the story they tell in one part of the world matches the story in another in a way that brooks no doubt. Second, the structure of the fossils and the strata in which geologists find them change slowly. On human time scales, they are effectively static—time capsules, if you like, though not deposited by builders in the cornerstone of a house for discovery in some future age. Nature put them there.

---

*Charles Darwin did not propose the idea of evolution. His grandfather Erasmus Darwin, along with others, had suggested evolution. The grandson's great insight was the mechanism. The three words "by natural selection", together with the simple explanation of what they mean, changed the world if not all minds. The culture wars in the United States revolve in great part about them.

Third, what they record is the growth of complexity. It is true that some species adapt to a changed environmental niche by becoming simpler, but such examples pale in comparison with the explosive growth of complexity from the unicellular Precambrian organisms to the human brain. The evolutionarily driven diversification of DNA tells an extraordinarily rich and consistent story of life's development on the earth. At the level of individual humans, this is helping to build up an intricate genealogical network of our ancestors. The skeleton of a hunchbacked human found buried a metre below a car park in Leicester could be unambiguously identified as the remains of Richard III because documentary and DNA evidence linked the skeleton to two living descendants of Richard through the female line. Records that can be especially trusted consist of diverse strands that have come down through history in different physical forms, often in locations that are far apart.

YOU PROBABLY DO not doubt the existence of the past; I didn't need to marshal the evidence for that in the preceding examples. What I want to emphasise is not the actuality of such evidence but what it tells us: that complexity is increasing in parallel with an ever greater mutual consistency of records. What we find on the earth, the heavens too proclaim. Consider stars. Most are long-lived; a detailed history of their formation and development has been established. Billions upon billions of them have clearly passed and are passing through a life cycle in which they develop an ever richer structure through the creation of atoms of ever increasing complexity—the stardust of which we are made. All this is now well understood on the basis of physical laws and observation. Astronomers' ability to understand these things now extends to the furthest reaches of the observable universe, both in space and in time. And all the evidence for this exists now, at this current cosmic epoch. In contrast to the stars, the development of galaxies is much less well understood. It's revealing that astronomers use stars as 'fossil records' and, seeking to understand the history of our galaxy, practise what they call galactic archaeology.

There is no need to multiply examples. We have learned many things about the world and the universe in which we live. One of the most remarkable is the cornucopia of records that, hanging together with great mutual consistency, tell a story of a universe of ever increasing complexity and structural richness. Appropriately analysed and interpreted, a grain of sand can now allow scientists to discern the history of the universe.

How do these stable records emerge, on both large and small scales, from the flux of history? How does it come about that the universe now is one huge time capsule of almost incomprehensible richness? I don't know the background to Muriel Rukeyser's motto in the epigraph to this chapter. Perhaps, like William Blake, she was reacting to the way science seems to drain joy and vitality from existence, replacing them with mere numbers. In fact, the billiard-ball-like atoms poets find so threatening are not at all like the atoms which quantum mechanics has discovered. In his *A Beautiful Question*, the physics Nobel laureate Frank Wilzcek holds that atoms are, literally, musical instruments.

By quoting Rukeyser, I don't in any way want to imply that atoms are the wrong basis for understanding the universe. We certainly cannot understand anything—our bodies, our brains, our phones, the aeroplanes we fly in—without the notion of atoms physicists now have. But what is the essential structure of any good story? It surely must have a beginning, an exposition, and an end. On that basis, the history of the universe is one huge story with countless diverse subplots all preserved in encoded form as records in the universe now. The universe is made of stories; scientists have learned how to decode and read them.

It turns out that they all unfold in a common direction. Bodley in Oxford is one of the world's great libraries, but we have been born into the ultimate library. It's all around us. It's the universe. One can argue, and I do, that the time-capsule nature of the universe—the way its state now proclaims a past history of growing complexity with extraordinary internal consistency—is the single most striking empirical fact that calls for scientific explanation. As of now, science can't provide it. All the known relevant laws of nature make no distinction whatsoever between the two possible directions of time. That's the rub. The laws tell us the timeline of the universe should be a two-lane highway. But all the stories I have mentioned—and many, many more—effectively erect a huge sign: THIS IS A ONE-WAY STREET.

I have presented this paean to the universe and its history not merely to emphasise the last point but rather as preparation for the claim this book makes. It is not just that it proposes a new theory of time and with it a first-principles explanation of its arrows; it also aims to overturn the doctrine that it is entropic disorder on a cosmic scale that puts the direction into time. The claim is this: the direction gets into time not through the growth of disorder but through the growth of structure and complexity. If that seems a rash claim, I invite you to go online and search for NASA's Timeline of the Universe. This and related images you will find at the same site show that at the big-bang birth of time there was a bland uniformity, from which there emerged, one by one, the most beautiful structures. In some you will see *Homo sapiens* as the current terminus; others show that terminus as the space probe WMAP, which a couple of decades ago made exquisitely accurate measurements of the very-low-frequency thermal radiation that bathes the universe and gives so many vital clues about our origin and place in the universe.

Suppose you show these images to the proverbial person in the street and say it's a story of the remorseless growth of disorder. I think the response might be 'Tell it to the marines'. If you consult a cosmologist interested in entropic questions, I think you will be told it's all something to do with the universe having a current entropy which is lower than the maximum it could have. This means the universe has an effective supply of free energy, which in thermodynamics as developed in the nineteenth century is the difference between the entropy that a system has at a given time and the maximum entropy it will have when no more useful work can be extracted from it. This is an account which is more or less compatible within conventional thermodynamics, though I think defining the entropy of an expanding universe as if the universe were a box of equilibrating gas is problematic to say the least. More seriously, I think that account shows only that entropy increase is not incompatible with the growth of structure. It does not show the direct connection between expansion of the universe and the way in which that expansion allows—indeed, mandates—the formation of specific structures. Above all, without the Janus-point idea it cannot explain the special initial condition that is needed, and therefore it cannot explain time's arrows.

CHAPTER 2

BARE-BONES THERMODYNAMICS

Heat and the First Law of Thermodynamics

In barely six pages, in which he proposed a method, albeit highly idealised, by which a steam engine could be made to operate with maximum efficiency, Carnot laid out all but one of the foundational principles of thermodynamics. His one mistake, which did not invalidate his proposal (known subsequently as the Carnot cycle), was to retain the contemporary theory of heat, which held it to be *caloric*, a weightless, incompressible, invisible fluid that could 'insinuate'—Maxwell's word—itself into bodies and cause them to expand. The mysterious substance came into physics in the late eighteenth century through a chance observation by Joseph Black, a professor at Edinburgh and adviser to whisky distillers. He happened to leave in a room two buckets; one contained ice and water, the other only water at its freezing point. A few hours later he found the second bucket's water noticeably warmer and in the first less ice and correspondingly more water but, to his surprise, still at the freezing point. Only when all the ice had melted did that water begin to get warmer.

Ambient heat in the room had obviously brought about the changes. But why had it only melted some of the ice in the first bucket without heating the water as well? It seemed heat could disappear. Believing its amount should remain constant, Black suggested it had become latent (Latin for 'hidden'). He also found that heating boiling water, while hastening evaporation, did not raise its temperature, which also suggested the heat had become hidden. To this day, students learn about latent heat, but now only as the heat needed to bring about phase transformations, such as the melting of ice or the boiling of water.

Carnot, for his part, likened the motive power of heat, presumed to be caloric, to that of water passing from a higher level to a lower one, driving a water mill in the process. In a steam engine, caloric would flow from the furnace at a high temperature into the cylinder, causing the steam in it to expand and move a piston. The cylinder would then be doused with cold water, which would take up caloric; with the caloric no longer present in the steam, the piston would return to the original position for the next cycle. Thus, according to Carnot, the motive power in steam engines is due "not to actual consumption of caloric, but to its transportation from a warm body to a cold body". Just as a waterfall carries water from a height to a depth, the steam engine carries heat in the form of caloric from a high temperature to a low one.

This all too plausible picture had nevertheless been effectively killed a quarter of a century earlier by the colourful and resourceful American-born British physicist and inventor Sir Benjamin Thompson, Count von Rumford, while supervising the boring of cannons in Munich for the ruler of Bavaria. He showed that an effectively inexhaustible amount of heat could be generated through the friction accompanying the boring. Bystanders were astonished "to see how, without fire, such a quantity of cold water could be heated and even brought to boiling".

The great question was therefore the one that "has so often occupied the natural philosophers: What is heat?" Having shown that the heat excited by the friction is inexhaustible and therefore could not possibly be a fixed amount of material substance, Rumford concluded: "It must therefore be *motion*". He published his findings in the august *Philosophical Transactions* of the Royal

Society of London and a chemistry journal in Germany. Despite this exposure, few scientists took any note of it for decades. The multitude of exciting experimental discoveries made with increasing frequency in the early decades of the nineteenth century is the probable reason.

IT WAS JAMES Joule who above all gave caloric the coup de grâce. He was a brewer in Manchester and, with every minute he could spare, an avid amateur scientist. He's famous for the simplest of experiments. He allowed the fall of weights to turn paddles in closed water-filled tanks. The churning of the water heated it by friction. Joule's brilliance was in measuring the tiny increase of the water temperature—perhaps the exigencies of brewing had something to do with that. By comparison, determining the work done by the turning paddles, measured by the total difference in height of the weights, was simplicity itself. This led Joule to the mechanical equivalent of heat: how much work must be done to generate a definite amount of heat. It is one of the most important numbers in science. In fact, what Rumford had measured in horsepower was, appropriately converted, reasonably close to Joule's pounds per foot fallen.

Joule's paper, published in 1843, attracted little attention until he spoke at the British Association in Oxford in 1848. The twenty-four-year-old William Thomson, already an expert in measuring very small temperature differences, was there standing at the back. He distrusted Joule's claim but, sufficiently intrigued, talked to Joule and corresponded with him. The result was prominent mention of the paddle experiment in Thomson's 1849 paper, the one that put Carnot's work on the map. Unfortunately for Thomson, he failed to register the critical importance of an earlier electrical experiment in which Joule had shown that heat could be transformed into mechanical work. Heat and work were completely interchangeable. Thomson later admitted somewhat ruefully in a letter to Joule his failure to appreciate that heat could not only be created but also be 'put out of existence' (by doing work). Failing to shake off Carnot's waterfall analogy, Thomson opined in his influential 1849 paper that the notion of caloric should not be abandoned before several more years of careful measurement.

CLAUSIUS HAD NO such inhibition. In his paper of 1850, he argued persuasively that a steam engine is not a water mill. The heat that comes from the furnace is not conserved; some is converted into work. It does not become hidden—latent—but ceases to exist; as in Joule's electrical experiment, nature has balanced things in the work done. The residual heat that remains in the working medium passes into the water sprayed onto the steam engine cylinder. Clausius's modification, and with it the completion of Carnot's theory, was the first clear example in which what is arguably the most fundamental physical law—the conservation of energy—is manifested. Thomson had missed that boat but soon still made an important contribution to the definitive formulation of the first law of thermodynamics by including in it the change in the internal energy of bodies. He was, in fact, the first person to establish widespread use of the word 'energy' and its two aspects: potential, as in the weight lifted or an apple on a tree, and kinetic, as the apple falls. The latter replaced the old name *vis viva* (living force). The difference between kinetic and potential energy in gravitational theory is going to be very important later in the book; it will be good to illustrate it by the example of a pendulum. When the bob is at its lowest point, its speed is greatest and the kinetic energy, which is always positive and equal to $mv^2/2$, where m is the mass of the bob and v is its speed, has its maximum value. At the highest point of its swing, the bob is at rest and has no kinetic energy. The bob's potential energy, for its part, is always negative, becoming less so with increasing height in such a way that throughout the motion the sum of the kinetic and potential energies remains exactly constant. As the pendulum swings back and forth, there is therefore a constant ebb and flow between the two forms of energy, subject to strict constancy of the total.

The discovery and formulation of the first law made a huge impression in what was still a profoundly religious age. The idea that God had created a universe in which, despite the vagaries of existence, something remains eternally unchanged was very comforting. Joule, as religious as Thomson, had already expressed the sentiment in 1847. As reported in the *Manchester Courier*, a talk he gave in that year brimmed with the confidence of a man who knows he had made a great discovery. Joule referred to one of the most important qualities with which God had endowed matter and that it would be absurd to suppose it could be destroyed or created by human agency.

Inviting his audience to behold "the wonderful arrangements of creation" and the vast variety of phenomena involving the conversion of living force and heat into one another, he said that they "speak in language which cannot be misunderstood of the wisdom and beneficence of the Great Architect of nature". It is something we see "in our own animal frames, 'fearfully and wonderfully made'". Indeed,

> the phenomena of nature, whether mechanical, chemical, or vital, consist almost entirely in a continual conversion of attraction through space, living force, and heat into one another. Thus it is that order is maintained in the universe—nothing is deranged, nothing ever lost, but the entire machinery, complicated as it is, works smoothly and harmoniously . . . the whole being governed by the sovereign will of God.

Coming back down to earth, it's a nice thought that caloric came into science through an adviser to distillers of whisky and went out through the efforts of a brewer of beer.

The Second Law, Dissipation of Energy, and Entropy

One of Carnot's inspirations was that heat was universal in its effect: whatever the working medium might be—steam or some simple gas—the efficiency of a steam engine operated under ideal circumstances would depend solely on the amount and temperature of the heat taken from the furnace and the amount and temperature at which it was cooled by dousing. He argued, by impeccable logic, that if this were not the case it would be possible to build a perpetual-motion machine from which one could extract an unlimited amount of work. But the claims of charlatans that they had created such machines had long since been debunked; it was now a well-established principle of mechanics that perpetual-motion machines could not be made.

Unfortunately there was a flaw in Carnot's proof—not in the logic but in the premise that caloric existed as an indestructible fluid. Carnot's idea impressed both Thomson and Clausius; recognising the flaw, both more or less simultaneously found marginal changes to his proof that rectified the false premise. Clausius was the first to publish a correct proof by noting that a thermal perpetual-motion machine could work only if heat flows spontaneously from a colder to a hotter body, which it never does. This simple observation was the first statement of the second law of thermodynamics. Clausius took pains to emphasise 'spontaneously' as critical.

Thomson, who fully recognised Clausius's priority in publication, adopted an alternative premise: "It is impossible, by means of inanimate material agency, to derive mechanical effect from any portion of matter by cooling it below the temperature of the coldest of the surrounding objects". He proved that Clausius's observation is equivalent to his premise, which is often illustrated by the fact that although the oceans contain a vast amount of thermal energy, that energy cannot be used to propel liners across them. The two laws of thermodynamics were soon seen to rest on the impossibility of constructing two different kinds of perpetual-motion machines: those of the first kind, which would give something for nothing, are ruled out by conservation of energy, while those of the second kind contradict the equivalent principles of Clausius and Thomson. Jokers like to call thermodynamics an example of 'sod's law': you can't get something for nothing and even if you have got something you can't use all of it. Quips aside, the anthropocentric reaction to thermodynamics and perpetual-motion machines, first expressed in Carnot's desire to improve the lot of humans through optimal use of coal, may be a distraction from the primary task: finding out what the universe is about. Unless we are gods, what it lets us do is secondary.

ALL THE EARLY rigorous results in thermodynamics were made using one of the two equivalent versions of the second law that Clausius and Thomson formulated in 1850. Clausius used his almost child-like remark to prove the existence of entropy and secure his place in the history of science. But already before that, in his influential 1852 paper 'On a universal tendency in nature to the dissipation of mechanical energy', Thomson, who always had an eye for engineering applications, had introduced one of the twenty or so alternative formulations of the second law that I mentioned in Chapter 1. The paper noted the wide significance of Carnot's criterion for operation of a steam engine with maximal efficiency: there will be an "absolute waste of mechanical energy available" to do useful work for humans if heat is allowed to pass from one body to another at a lower temperature without any work being done. Being convinced that "Creative Power alone can either call into existence or annihilate mechanical energy", Thomson said the 'waste' could not be annihilation but must be some transformation of energy, for which he chose the word 'dissipation'.

To clarify what he meant, Thomson defined the notion of stores of mechanical energy. In one broad class he included static examples such as weights at a height, electrified bodies, and quantities of fuel; in a second class are dynamical stores, as in the case of masses in motion, a volume of space through which undulations of light or radiant heat are passing, or a body having thermal motions among its particles. The two classes match the two forms of energy—potential and kinetic—whose names he did so much to establish.

One among several examples of energy transformation he gave is that when heat is created by any irreversible process, such as through friction, "there is dissipation of mechanical energy, and a full restoration of it to its primitive condition is impossible". The implication of this and other examples was that "within a finite period of time past the earth must have been, and within a finite period of time to come the earth must again be, unfit for the habitation of man". This was inevitable, according to Thomson, "under the laws to which the known operations going on at present in the material world are subject".

Thomson, a prolific author who tended to dash off influential papers without details his readers might have appreciated, did not make explicit the reasons for the finite duration of possible past and future life on the earth. However, for readers who had grasped the significance of the conservation of energy expressed in the first law of thermodynamics the reason for the termination of terrestrial life could be guessed. The sun and the earth have only finite stores of energy and both are cooling by radiation. The second law, expressed as dissipation, then delivers the killing blow, making it possible to predict with confidence the future of the solar system in broad outline. It must be one of progressive cooling, leading to the end of the world as a habitation for humankind—heat death, as it was soon called. If conservation of energy was widely seen as a reassuring aspect of thermodynamics, confirming Joule's enthusiasm, the prospect of heat death cast a pall that has never gone away.

As for the finite duration of life in the past, Thomson made this clear in 1854 at a meeting of the British Association at which he stated confidently that the ultimate source of all mechanical energy in the solar system must be gravitational potential energy of a primordial nebula that, by our epoch, had been largely transformed into hot bodies, either solid or liquid. Although Thomson of course had no idea that it is nuclear and not gravitational energy that plays the overwhelmingly dominant role in the sun's energy balance, granting it a life span beyond anything he could imagine, his broad outline of the origin and evolution of the solar system possessed of only a finite amount of energy and subject to the laws of thermodynamics is in essence correct.* However, beyond

---

*Over decades Thomson and his friend Peter Guthrie Tait engaged in increasingly testy exchanges with geologists and Charles Darwin's supporter Thomas Huxley over the great age of the earth that both geology and the theory of evolution posited; since the sun and earth could have only a finite store of energy, the laws of thermodynamics seemed to rule out such an age. Huxley became known as 'Darwin's bulldog'; Tait played the same role for Thomson. It was a proper dogfight. Darwin was more than a little discomfited by the attacks of Thomson, who was widely regarded as the greatest British scientist of the age, not only as joint discoverer of the two laws of thermodynamics but also for many other achievements. He was ennobled as Lord Kelvin in 1892, and after his death in 1907 he was buried with great pomp next to Isaac Newton in Westminster Abbey.

an act of God, Thomson could give no ultimate explanation for the existence of any of his stores of mechanical energy, including the nebula he supposed had given rise to the solar system. Dissipation of mechanical energy is undoubtedly a universal tendency in nature—Thomson was right about that—but ideally a complete law of things should explain how the stores arise in the first place.

CLAUSIUS'S DISCOVERY OF entropy, using simple but subtle arguments, is a wonder. He found a new state function. What kind of beast, you may ask, is that? The best modern account of it, its discovery, and the way to prove it exists is in the Dover edition of *Thermodynamics* by the Nobel laureate Enrico Fermi, given as lectures in 1935.* However, it's a tightly argued account with plenty of equations, so here, with as few of them as possible, I outline what Clausius did, beginning with basics. I think it will help if you try to get to grips with at least the most important points; I will explain why at the end of the chapter.

Hopefully, my nonscientific readers remember from the start of the book the importance of equilibration and that it always leads to equilibrium. When that is achieved, say for a gas in a container of volume V, then the pressure P and temperature T are constant everywhere and, if the gas is sufficiently far from its condensation temperature, the ideal-gas law $PV = NRT$ holds (N is the mass of gas and R, which has the same value for all gases, is called the universal gas constant). P, V, and T are state functions; they characterise the equilibrium state. Any two fix the third and with it the equilibrium state. Temperature is a particularly interesting state function. Since the early eighteenth century the ideal-gas law had suggested something special must happen at about –273°C. This became known as the absolute zero of temperature through the theoretical work of Thomson, who, building on Carnot's insights, showed that the mechanical effect of heat could be used to measure temperature. This eliminated the problem that the thermometers hitherto used, which were based on expansion of gases and

* It can be found as a PDF online to be downloaded free. You might look at it briefly just to get an idea of the ingenuity that is involved. For physics students, I recommend it highly. I am grateful to Domenico Giulini for recommending Fermi's booklet.

liquids, did not give concordant results. Thomson's measurements also determined the absolute zero of temperature with great accuracy, in recognition of which absolute temperatures are measured in kelvins.

IF CALORIC DID exist and could be neither created nor destroyed, it too would, like pressure and temperature, be a state function. But Joule's experiments had done away with caloric. Already in 1850 Clausius began to suspect some other state function might take its place. By 1854 he had it.

Carnot's idealised steam engine required heat to be transferred from a furnace to steam in a cylinder at a temperature only infinitesimally lower. The steam would be allowed to expand very gradually, thereby raising infinitesimal weights that could be removed one by one at successive heights. In the ideal limit, this process is reversible: one by one the weights are put back on the piston, causing it to sink and raise the temperature of the steam ever so slightly above that of the furnace, into which heat would now flow. Were heat to be transferred from the furnace with finite temperature difference, some of the heat would simply raise the temperature of the steam without being fully exploited to do work. As Thomson had noted in 1852, Carnot placed great stress on the need to ensure that the maximum possible amount of work should be done whenever heat passed from a higher temperature to a lower one. Water mills worked on the same principle—no water should fall without doing some work.

The identification of reversible transformations was one of Carnot's several great contributions to science. Clausius took his work further into a quite new realm. His great idea was that one could gently 'nudge' enclosed gas from one equilibrium state to another by letting it exchange infinitesimal amounts of heat with an adjacent heat reservoir at an infinitesimally higher or lower temperature. This insight came straight from Carnot.

Here I must introduce two or three of the symbols that Clausius employed; they have become standard. By Q he denoted an amount of heat and by dQ an infinitesimal amount of it. A reversible transformation then involves either the infinitesimal transfer of dQ of heat to the gas or –dQ from it. The decisive step was that Clausius then divided the amount of heat dQ by the absolute temperature T and called the resulting expression

$$\mathrm{d}S = \frac{\mathrm{d}Q}{T}$$

the (infinitesimal) transformation value of the process. It was only in 1865 that, with aplomb, Clausius replaced the clumsy phrase 'transformation value' by 'entropy', which he still denoted by S and the infinitesimal increment of entropy by dS. It's possible that, to honour Carnot, he chose S for Carnot's first name, Sadi. It's a nice idea.

Your next question might be to know the significance of the division of the amount of heat dQ by the absolute temperature T. Well, Joule's experiments had shown that all heat can do work, so dQ could, say, lift eight ounces through ten inches. Now temperature measures something like the 'quality' of heat. You feel that when coming from the cold into a warm room. In fact, if you are feeling cold and have the option of going into one of two rooms, one large and one small and both containing the same amount of heat, you will surely go into the small and hotter room.

Reflecting that, the increment of entropy dS in Clausius's definition measures the quality of the added heat dQ. If it is added at a low temperature T, then the entropy increment dS = dQ/T is large; the heat is, as it were, 'spread out' or diffuse, as indeed it is in the large room you would not choose to enter. If T is large, dS is small, the heat is more 'concentrated' and it has higher quality. You can achieve more with it.

I MUST NOT omit to illustrate here this matter of the quality of heat by the most important engineering insight that emerged through the work of Carnot, Thomson, and Clausius. The Frenchman had proved that the maximum possible efficiency of a steam engine depends solely on the two temperatures between which it operates:

the temperature of the steam heated by the furnace, T_{steam}, and the temperature of the steam-dousing coolant, $T_{coolant}$. However, he had not been able to find an expression for the maximum possible efficiency. Mainly through Thomson's definition of the absolute scale of temperature, it was shown to be the difference between these two temperatures divided by the absolute temperature T_{steam} of the steam. This then is the great triumph of thermodynamics: the maximum possible efficiency of a steam engine, typically denoted by the Greek letter η (eta), is

$$\eta_{\text{maximum}} = \frac{T_{\text{steam}} - T_{\text{coolant}}}{T_{\text{steam}}}.$$

Unless the steam is superheated, its temperature is 373 kelvin, while the coolant will typically be at the ambient temperature, say 293 kelvin. This makes the maximum efficiency

$$\eta_{\text{maximum}} = \frac{373 - 293}{373},$$

which is only a bit over 21 per cent. Most steam engines work with a much lower efficiency. The very first ones achieved barely 2 per cent. You might like to go online to read about the first commercially successful steam engine invented by Thomas Newcomen in 1711 to pump water from mines. Note that the formula for the efficiency expresses it as a ratio. This exemplifies the fact, noted in Chapter 1, that it is only ratios that have physical significance.

SO FAR I have only told you how Clausius defined the increment of entropy; I have not described how he defined entropy itself or what he considered to be its full significance. Let's start with the definition. Clausius supposed some closed system, for example ideal gas in a cylinder whose volume can be changed by a piston. The gas is initially in some equilibrium state, to which Clausius ascribed a nominal entropy $S_{initial}$. He then imagined the gas taken by infinitesimal reversible transformations through a succession of equilibrium states, each with its own pressure, temperature, and volume.

The only thing he took into account was the heat taken from or given up to a series of heat reservoirs by the gas at each stage of his delicate manoeuvre. Any work the gas might do through the piston or the piston might do to it did not enter his calculations. He was solely concerned with entropy increments d*S*. They should be added to (for a positive amount of heat, + d*Q*) or subtracted from (for a negative amount, –d*Q*) the current value. Then after *N* steps he arrived at a final value of the entropy:

$$S_{\text{final}} = S_{\text{initial}} + \mathrm{d}S_1 + \mathrm{d}S_2 + \ldots\ \mathrm{d}S_N.$$

Clausius proved that no matter which of the infinitely many 'paths' through the possible equilibrium states was chosen, S_{final} would always be the same and depend only on the initial and final equilibrium states. If they, along with the nominal value of S_{initial}, are specified, then the difference $S_{\text{final}} - S_{\text{initial}}$, and with it S_{final}, is fixed. It is path-independent.

The notion of path independence is one of the most important in both mathematics, the queen of the sciences, and physics. I therefore owe you at least some kind of explanation of what it means. Luckily it's not too difficult. Consider first the surface of the earth. At every location in a given country, the height above mean sea level, the altitude, has a fixed value. If you go by any path from one given location to another, your altitude will change by the same amount whatever route you take. The difference in altitude is path independent. But now suppose the country has toll roads with variable rates, and on some you make payments but on others you are given bonus payments to encourage travel along routes which do less harm to the environment. Then obviously the cost of a journey will depend on the route you take. It will be path dependent. I think you will agree that the basic idea is simple even if its applications can be very subtle.

It was in Clausius's case. His proof that the difference between S_{final} and S_{initial} depends only on the initial and final equilibrium states and not on the path taken between them is extraordinarily clever and involves heat transfer not only to or from the *N* heat reservoirs but also to and from just as many heat engines. Examples of these are practical things like refrigerators, the first 'thermodynamic spinoffs', brilliantly exploited by Thomson wearing his engineer's hat in the Joule-Kelvin process, as well as Carnot's model of an idealised steam engine working at maximum efficiency. Besides all this 'auxiliary machinery', Clausius employed one further heat source at a uniform fixed temperature. It is there so that, at the critical point in the argument, it can be argued that if S_{final} were not path independent, work would have been done in violation of the second law: by extracting heat from a uniform source (Thomson's form of the second law as used by Fermi) or, as in Clausius's original proof, heat would have flowed spontaneously from a colder body to a hotter one. The power of the second law in either of these utterly simple formulations is breathtaking. Fermi uses it twice to establish the absolute thermodynamic scale of temperature and once more to prove the path independence of S_{final}.

This is what enabled Clausius to argue that S_{final}, not yet called entropy, is a state function. Its dependence on S_{initial}, which can be fixed once and for all by choosing a standard reference state, is of no concern. The real thing of interest was the change, the resulting difference, in going from one state to another. What Clausius had found is a quantity that characterises the capacity of a considered substance to do useful work. The temperature tells you what it feels like; the new state function tells you something about what you can do with it. I should also say that although the nominal value given to S_{initial} is immaterial, some reference state and value must be chosen.

IN A LATER paper, of 1862, Clausius went further in drawing attention to a result he had already known in 1854 but had withheld because he considered it too controversial for his peers in the way it overturned then-current beliefs. To establish the 'disturbing' result, he relaxed the condition that all heat transfers must be made reversibly. He now allowed heat to flow between an enclosed system and heat reservoirs with a finite and not infinitesimal temperature difference. This was the adverse possibility that had led to Thomson's comment about the 'absolute waste' of mechanical energy available to humans and the notion of heat death. Clausius, for his part, showed what happens if gas is taken around a 'closed loop' of heat transfers with heat reservoirs at temperatures differing by a finite amount from the gas. He quantified the outcome by the expression

$$\oint \frac{\mathrm{d}Q}{T},$$

where the rather beautiful integral sign $\oint$ invented by mathematicians means one is adding up (integrating) quantities around a closed loop. It is important, as Fermi points out in a footnote, that T is the temperature of the source (the system or a heat reservoir) that is giving up the heat. Under these conditions it is possible to show that if the above integral were greater than zero, work would have been done in violation of the condition that heat cannot flow from a colder body to a hotter one (or work done by heat received from the further source at uniform temperature, which here plays a critical role). Thus, the integral must be less than or equal to zero. If one now supposes that all the transfers are reversible and are made in the opposite direction, one can conclude that the integral must be less than or equal to zero. From this it follows that in a reversible process the integral is zero,

$$\oint \frac{\mathrm{d}Q}{T} = 0,$$

which, the integral being Clausius's definition of the entropy based on reversible processes, confirms that entropy is a state function defined in a path-independent manner, since its value is unchanged when the system comes back to its original equilibrium state. However, if the heat transfers are made irreversibly one has the inequality

$$\oint \frac{\mathrm{d}Q}{T} < 0.$$

For readers who consult Fermi's text, they will find that he proves the existence of entropy as a state function simultaneously with showing that it can never decrease in an isolated system; in contrast, Clausius did this in two stages separated by eight years. For this reason, Fermi, who wants a conclusion that will cover both reversible and irreversible transformations, retains the possibility that the integral may be either less than or equal to zero but never positive.

It may help to explain why the above inequality holds. As Fermi pointed out, what counts in determining the value of dQ/T is the temperature of the source that gives up the heat, not the temperature of the source that receives it. Now suppose that at some stage the system gives up heat to a reservoir which has a temperature that is lower by a finite amount and not infinitesimally. Then dQ/T, which will be negative since it is the system that is giving up heat, will be larger in magnitude than it would be in a reversible transfer. If, on the contrary, the system takes up heat from a reservoir at a temperature higher by a finite amount, dQ/T will be positive but smaller in magnitude than it would be in a reversible transfer. Thus, either possibility that allows heat transfer at a finite temperature difference will make the integral around the loop negative.

Since the integral we have considered looks like Clausius's expression for the change in the entropy, it might seem from the conclusion just drawn that the entropy will *decrease* in an irreversible transformation. However, this is not so because we are considering different things. In Clausius's definition of entropy we pass from one equilibrium state to a different one by reversible transformations through a succession of equilibrium states. In this case we are coming back to the same state and the system has been involved in at least some irreversible heat transfers. If one considers an irreversible transformation from an initial equilibrium state A to another equilibrium state B followed by a reversible transformation back to the original equilibrium state A (which can be brought about by letting the system expand or be compressed by a piston without any heat transfer), one can, using the inequality proved for the integral around the closed loop, directly show that the entropy must have increased in the passage from *A* to *B*. This is spelled out in Fermi's booklet.

In a final step one can show that for a system that is thermally isolated, so that no heat transfer from any external heat source is possible, the value of S_{final} reached by transformations within the system can only be equal to or greater than S_{initial}. The modern statement of this fact is that the entropy of an isolated system cannot decrease: it either remains constant or increases. A very simple example of this is realised when inside a closed insulated system there are two compartments of gas, initially at different temperatures, separated by an insulating partition.* In this state, the temperature difference could be exploited to do work, but if the partition is removed the temperature is equalised and that opportunity is lost. This is an example of the fact that you cannot use the thermal energy in the oceans to propel liners across them.

For what will be discussed shortly, Clausius's definition of the entropy of a system whose temperature is not uniform throughout is important. He said that in such a case "the expression for the entropy must not be referred to the entire body, but only to a portion whose temperature may be considered as the same throughout; so that if the temperature of the body varies continuously, the number of parts must be assumed as infinite. In integrating, the expressions which apply to the separate parts may be united again to a single expression for the whole body". In contrast, Fermi is much more rigorous and defines entropy in such a case only for a system composed of several homogeneous parts at different temperatures and pressures but with each "in a thermally insulating rigid container"; then the "system will be in equilibrium, and we shall be able to determine its entropy".

---

* Those of you who do read Fermi's excellent booklet will find, in what seems to me a curious lapse, that he gives the statement of the previous sentence without a proper proof. All his previous arguments rely on transfers with external heat sources. I do not see how any conclusion can be drawn if there are none. The example I give with removal of the partition makes the conclusion intuitively obvious.

As Clausius was breaking new ground in 1862, it's not surprising that he did not anticipate the thermally insulating rigid container that Fermi requires for each homogeneous part. To a significant extent, rigorous thermodynamics treats highly idealised situations. This is seldom a problem. In a laboratory, it is possible to construct containers that, for the states of matter being studied, are to an excellent approximation thermally insulating and rigid. But we do not see that sort of thing around us in the universe at large. And we do not see it in Clausius's words either. This must raise doubts about that part of his arguments and what followed from them. In fact, when I come to describe the evolution of the universe in Chapter 19 I will rely much more on Thomson's qualitative dissipation of mechanical energy, which can always be recognised, rather than entropy, which, as we have seen, can only be properly defined under very special circumstances.

IN HIS HIGHLY influential paper of 1865, which completed his series of papers on thermodynamics, Clausius introduced his alternative to the rather bland 'transformation value'. He clearly did this to alert people to the full significance of the concept. In introducing his coinage, Clausius said he held it to be better "to borrow terms for important magnitudes from the ancient languages, so that they may be adopted unchanged in all modern languages". He chose 'entropy' from the Greek word τροπη, 'transformation', adding to it the prefix 'en', saying, "I have intentionally formed the word *entropy* so as to be as similar as possible to the word *energy*; for the two magnitudes to be denoted by these words are so nearly allied in their physical significances that a certain similarity in designation appears to be desirable".

The passage prepares the way for the dramatic end of the paper. Clausius says he wants to allude, at least briefly, to a subject of which "even a brief statement may not be without interest, inasmuch as it will help to show the general importance of the magnitudes which I have introduced". The second law, he says, asserts that all transformations in nature may take place in one direction spontaneously but not in the other. He says this "leads to a conclusion to which W. Thomson [in the 1852 paper] first drew attention". Clausius ends the paper with a flourish:

> If for the entire universe we conceive the same magnitude to be determined, consistently and with due regard to all circumstances, which for a single body I have called *entropy*, and if at the same time we introduce the other and simpler conception of energy, we may express in the following manner the fundamental laws of the universe which correspond to the two fundamental theorems of the mechanical theory of heat.
>
> 1. *The energy of the universe is constant.*
> 2. *The entropy of the universe tends to a maximum.*

I have not found anything in Clausius's publications that warrants the second statement. There is no trace of the rigour of his great definition of entropy for enclosed systems that pass under tightly controlled conditions from one state to another. All we have is the brief statement of 1862 on how entropy is to be defined for a system that is not at a uniform temperature throughout. Fermi's conditions show how careful one must be in defining entropy for a system in an inhomogeneous state. There is no trace of that in the buildup to what Clausius clearly saw as an epochal conclusion. Nevertheless, the 'fundamental laws of the universe' had, and still do have, a great effect both among the lay public with interest in science and among working scientists. An example of the latter was J. Willard Gibbs (the greatest American scientist according to Einstein), who in 1873 published a very important paper that extended notions of thermodynamics to different chemical substances. As the epigraph at the head of his paper, he quoted Clausius's two 'fundamental laws of the universe'.

Because the two laws of thermodynamics seemed to rest on such secure foundations—the impossibility of building perpetual-motion machines of either the first or the second kind—they soon acquired a well-nigh inviolable status. Near the end of his life,

Einstein said of thermodynamics: "It is the only physical theory of universal content which I am convinced that, within the framework of applicability of its basic concepts, will never be overthrown". In lectures given in 1927, Arthur Eddington, who, as noted, coined the phrase 'arrow of time', said that the law

> that entropy always increases holds, I think, the supreme position among the laws of Nature. If someone points out to you that your pet theory of the universe is in disagreement with Maxwell's equations—then so much the worse for Maxwell's equations. If it is found to be contradicted by observation—well, these experimentalists do bungle things sometimes. But if your theory is found to be against the Second Law of Thermodynamics I can give you no hope; there is nothing for it but to collapse in deepest humiliation.

However, Einstein cautioned that the universality of thermodynamics holds only within the framework of applicability of its basic concepts. In 1865, Clausius had not the remotest idea that the universe is expanding; in 1927 Eddington was just beginning to be aware of the possibility. Clausius did rather blithely assume that defining the entropy of the universe would be relatively straightforward. There may be a devil in the details of his phrase "consistently and with due regard to all circumstances".

CHAPTER 3

STATISTICAL MECHANICS IN A NUTSHELL

BESIDES THE IDEAL OF BREVITY, A NUTSHELL REFLECTS THE NEED FOR CONfinement that the steam engine brought into thermodynamics. Hamlet said, "I could be bounded in a nutshell, and count myself a king of infinite space, were it not that I have bad dreams". The bad dream in thermodynamics—heat death—certainly comes from confinement. Gas in a box must equilibrate; there is no evading the end state. What, if any, is the deep reason for it?

It was sought through the atomic hypothesis, which served as the basis of a microscopic explanation of the great discoveries of thermodynamics. They had been found entirely through macroscopic measurement of effects one can literally see and feel. The aim then became to explain the discoveries of this phenomenological thermodynamics through the action of invisible atoms and molecules subject to mechanical laws. This led to the creation of statistical mechanics, which combined Newton's laws of mechanics with statistical arguments made necessary by the vast number of atoms in even a very small amount of gas. There was no way the motion of each individual particle could be followed. Among the triumphs mentioned in Chapter 1—above all the proof that atoms and molecules of definite sizes exist and the first discovery of quantum effects—a high point was Boltzmann's remarkably simple microscopic explanation of Clausius's visually beautiful but conceptually enigmatic

$$\oint \frac{dQ}{T} < 0.$$

That explanation is going to be the main topic of this chapter.

The basic ideas could hardly be simpler. In Newtonian mechanics, the notion of a mass point had long been established. Its sole attribute besides being a point was to have a mass, a positive number that could in principle have any value.* However, I need to recall here what I said in Chapter 1 and have mentioned a couple of times since then: in physics, all meaningful statements are about ratios. Whether measured in inches or centimetres, my little fingers are 1/3 the length of their hands. Subject to the accuracy of my measurement, that's an objective fact. Correspondingly, only mass ratios have meaning; physicists have various ways to determine them. But now you might see a difficulty with points, which have no length. It is clearly impossible to form size ratios for them. However, in mechanics mass points are assumed to exert forces that have a characteristic range r. For two mass points separated by distance d, the ratio r/d is meaningful. This justified the use of mass points as models. In fact, the founders of statistical mechanics obtained very similar results when they modelled gases either with mass points that exerted forces of some finite range or elastic billiard ball models of fixed size in the same range. When serious mechanics-based theorising began in the 1850s, it was also important that chemists had already provided good evidence for the existence of molecules and shown that they must consist of two or more atoms bound very close to each other.

* Mass and weight are different but, as Newton was the first to realize, they are always exactly proportional to each other for all bodies. That's why a hammer and feather, dropped on the airless moon, fell at the same rate. Einstein made the exact proportionality the cornerstone of his general theory of relativity, calling the simple realisation "the happiest thought of my life".

Under the assumption that atoms or molecules obey Newton's laws, moving mostly along straight lines at uniform speed except when encountering other atoms or molecules or the walls of a confining box, a broad intuitive picture of equilibration is readily obtained. Consider a large number of atoms and suppose that, at some initial time, they are all for some reason concentrated in one corner of the box and are moving in various directions with a certain range of speeds. They will soon spread out within the box, and it is easy to imagine that as a result of collisions with each other and the box wall a more or less uniform distribution of the atoms will be established throughout the box. One can also expect that some stable distribution of the atoms' speeds will be established.

ALTHOUGH OTHERS HAD already drawn interesting conclusions from such a picture, above all an explanation of pressure, Clausius was the first, in a paper in 1856, to make real progress. He obtained several key results that have stood the test of time. In particular he proved what is known as the law of energy equipartition: in an equilibrium state particles of different masses must, within a certain spread, all have the same mean kinetic energy. Since this is $mv^2/2$, where v is the particle's speed and m its mass (the division by 2 is introduced purely to simplify other expressions), this means the more massive particles move slower than the less massive ones. However, Clausius made no attempt to determine what the spread about the mean would be.

Fortunately for the development of science, his paper, together with one that followed it, attracted the attention of the brilliant James Clerk Maxwell, whose greatest achievement was to discover the laws of electromagnetism. In 1859 he published a paper in which he gave a simple intuitive argument for the distribution of the velocities, and equivalently the kinetic energies, of gas particles when in equilibrium. He predicted that for particles in one dimension the energies should be distributed as in the famous bell curve, or rather its positive half since the square of the speed in the expression for the kinetic energy necessarily makes the energy positive.

Examples of the form the distributions take in three dimensions are shown in Fig. 1. They correspond to different total energies of the particles, while the average energy determines the temperature in the equilibrium state. It is important that in statistical mechanics temperature is a quantity that can be defined only for a system in equilibrium. This corresponds to the fact that temperature, like pressure, can only be measured once a gas has equilibrated.*

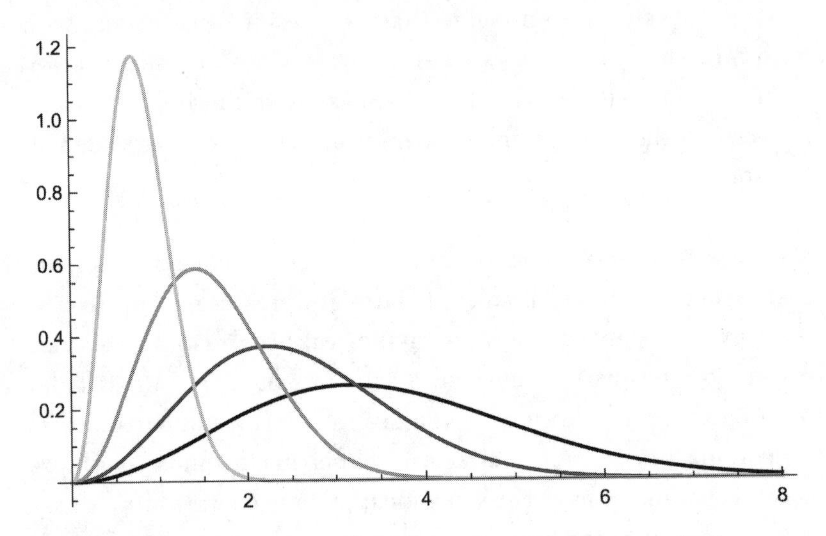

FIGURE 1. Maxwellian velocity distributions for a monatomic gas in three dimensions.

Maxwell relatively soon came to doubt the soundness of his derivation; in a most important paper in 1867, in which he greatly extended the scope of statistical mechanics, he said people might find his original proof 'precarious' and replaced it by another based on detailed calculations of the effects of collisions between the particles of the gas. Together with a generalisation by Boltzmann to

*Although the air around us is not in a physical box, it is effectively confined by the gravitational field of the earth and is, to a good approximation, in equilibrium at different altitudes. In fact, if enclosed in an impermeable, perfectly insulating column, it would have the same temperature at all heights (but different density). This extension to Maxwell's result was one of Boltzmann's first predictions.

include not only kinetic but also potential energy the distribution soon became what it still is: one of the most important relations in physics.

However, Maxwell only aimed to find the distribution and to show that once a gas was in the equilibrium state, it would remain in that state. He made no attempt to explain how equilibrium could arise from a non-equilibrium state; it was Boltzmann who undertook that task. Expressed in the simple terms that I have just used, it might seem to be a relatively simple matter. Does the behaviour of gas in a box bear much relation to the great scheme of things? It does. If the all-powerful laws of nature, long established by Newton, make no distinction between the two possible directions of time, how does it come about that all observed processes, including mundane equilibration in a box, do? Boltzmann attacked the problem head-on. Whether he fully appreciated the magnitude of the undertaking he commenced in 1872 is not clear. What is clear is that it soon became the defining issue of his life. Through it he became justly famous.

HIS BASIC TOOL was just the same as Maxwell's—to study the effect of collisions—but with a difference: not to see how they would maintain an equilibrium state but to see how they would change one not in equilibrium to one that is. He simplified his task by two assumptions. First was an assumption that, as I already said, seemed obvious: a gas that initially has a non-uniform spatial distribution in a box would relatively soon spread out into a uniform distribution, with the density effectively the same everywhere. By the time Boltzmann was working on the problem, everyone had a pretty firm conviction that Avogadro's number, essentially the number of atoms or molecules in a gram of substance, was huge. Its value depends on the kind of particles that make up the substance, but is typically of the order of 10^{23} (getting on for a million million million millions), so Boltzmann could safely assume that even in a very small region the density of particles would vary very little from point to point once a uniform distribution was established. He also made the plausible assumption that at any point the particles would be moving with equal probability in all three spatial directions.

Much less obvious was how the energy distribution could get to be Maxwellian. After all, the curves in Fig. 1 have a very specific form; anyone could expect a uniform spatial distribution to arise, but how could the kinetic energies get to be described by the curves shown in the figure? This was the problem Boltzmann addressed in 1872. He first defined a function, later called the *H* function, that was a measure of the difference between a current changing distribution of kinetic energies and the Maxwell distribution. His task was then to show that collisions between the particles of the gas would progressively reduce the difference between the two. To prove this, he derived an equation that reflected the way collisions would change the *H* function. It is now called the Boltzmann equation and is one of the most important equations in physics, being used to describe how distributions change in the most varied of situations—from fluids in small flasks to matter distributions in cosmology.

At the level of the distribution of the kinetic energies of the atoms or molecules of a gas, Boltzmann was able to show in his 1872 paper that if the distribution did change in accordance with his equation the *H* function must decrease, and therefore there must be a tendency for establishment of the Maxwell distribution. This result, which required insight and sophisticated mathematics, was no mean feat. Boltzmann felt he could rest on his laurels.

BUT ONLY FOR a few years. In 1877, his friend Joseph Loschmidt published a paper that cast doubt on Boltzmann's 1872 paper.* He did not do this in any way to undermine the main thrust of Boltzmann's work; in fact, he hoped it might help to "destroy the terrifying nimbus of the second law, which makes it appear to be a destructive principle for the totality of life in the entire universe". Loschmidt's

---

* Loschmidt, a colleague at Vienna, had used a result from Maxwell's study of viscosity to obtain the first estimate of the size of molecules and, with it, what English texts call the Avogadro number. In German texts it is, understandably, called the Loschmidt number.

argument was simplicity itself: time-reversal symmetric mechanical laws had been used to obtain a manifestly time-asymmetric result. There must be a flaw somewhere. Discussions about the issue rumble on to this day, but I am not going to get into them since they are technical and Loschmidt's intervention prompted Boltzmann to a change of tack. It led to a wonderfully transparent explanation of equilibration and an explicit microscopic expression for entropy and its increase that Clausius, in his phenomenological treatment, had expressed through the enigmatic relation repeated at the start of this chapter.

In the very same year as Loschmidt's paper, Boltzmann published two papers. In the first, which was discursive with a minimal number of calculations, he granted the validity of Loschmidt's objection but emphasised—as he had not done in 1872—that an atomic explanation of the second law could not be based purely on mechanical laws but must take into account statistical considerations. This involved considering all possible states that a large collection of particles could have. The definition of any one state would then consist of all the positions of the particles, any one set of which defines a possible *configuration*, while the collection of all possible configurations is the *configuration space*. There is a companion space for the velocities or, more fundamentally from the mechanical perspective, their momenta (the momentum of a particle is the product of its mass and its velocity, which is defined by its speed and direction). The space of all possible sets of momenta is called the *momentum space*. Boltzmann called the state that is completely defined by giving all the particle positions and momenta the *phase* of the system, and he applied statistical arguments to all of its possible phases. In fact, the collection of such phases for any given system is its most fundamental mechanical characterisation, but it was only as late as 1911 that, almost incidentally, it was called *phase space*. If you can, please remember the concept and its name; it will occupy centre stage in Chapter 4, where I will try to give you a better idea of what it is like.

What is important now is that Boltzmann had the insight that "there are many more uniform than non-uniform states". In the second 1877 paper he spelled out what this meant and its implications. They are often illustrated, in two dimensions, by the number of ways in which a certain number of chequers can be placed on a given number of chequer squares. I will make the illustration even simpler by using a chequerboard in the form of a ring divided into seven bins numbered 1, 2, 3, 4, 5, 6, 7. Suppose there are seven chequers *a*, *b*, *c*, *d*, *e*, *f*, *g* each assumed to have their own individual identity like you and me. Let's think about the numbers of different ways the bins could be filled with chequers. They could, for example, all be put in one bin, creating a 'tower'. Since there are seven bins, there are seven ways of doing that; all of them correspond to a very non-uniform distribution. At the opposite extreme, you could place one chequer in each bin. There are many more ways of doing that. Chequer *a* could go in any one of the seven bins, chequer *b* in any one of the remaining six, and so forth. That's $7 \times 6 \times 5 \times 4 \times 3 \times 2 \times 1 = 5040$ possibilities and 720 times more than the 'single tower' possibility. Now let's see what happens if we make a tower of six chequers in one bin and place the last chequer in some other bin. The tower can be in one of seven bins, that's seven possibilities, and the odd chequer left out of the tower in one of the six remaining bins. That's 42 possibilities. You can see already how the uniform distributions greatly exceed the non-uniform ones.

Thus, finding the numbers amounts to finding all possible permutations; it is a well-defined mathematical problem in statistics entirely unrelated to the laws of mechanics. The key point Boltzmann realised is that if the number of chequers is much greater than the number of bins, the greatest number of possible distributions is realised in the example when the heights of the 'towers' are all much the same. It's no good having a distribution like the skyscrapers of Manhattan, with a wide range of heights and here or there a few one- or two-storey dwellings. You want to make the distributions as uniform as possible. In fact, once you start to calculate the number of possibilities with any sizeable number of chequers and squares, you find that the number of possibilities with the most uniform 'skylines' vastly exceeds not only all the others but also all of them added together. As we will see, entropy is going to be defined in terms of the number of possibilities. It is often said to be a measure of disorder; this is perfectly true at the microscopic level because the chequers can be muddled up every which way. However, the macroscopic distribution that then results is the one that is most uniform. Thus, microscopic disorder corresponds to macroscopic order.

It will be helpful here to introduce two key concepts that underlie Boltzmann's statistical considerations. The first is that of a *macrostate*. In the example with chequers in bins, the macrostate is defined by the numbers of chequers in the individual bins. Thus, the fact that all chequers are in bin 7 defines one possible macrostate; so does the distribution with six in bin 7 and one in bin 4. The possible macrostates are defined without regard to which chequers are in which bins. Manhattan is again a good illustration; the outline you see from a distance defines its macrostate. If the skyscrapers were all built of numbered bricks, the *microstate* would tell you which bricks had been used in the construction of each skyscraper.

Before concluding this discussion of spatial distributions, I should mention that in mechanics the particles can be anywhere with definite positions and not just within certain squares. The introduction of bins in which you imagine the particles being placed is called coarse graining and does give rise to some muttering among purists. Provided there are enough particles to ensure that average densities change little unless the bins become very small, I don't think it's a serious problem; I will say something about the issue in Chapter 4, in which it becomes important for a different reason.

IN CONNECTION WITH this, and before I proceed to Boltzmann's treatment of kinetic energies, I want to go back to his 1872 paper, where he first introduced coarse graining. He did this by assuming that any particle could only have discrete multiples of a given small energy ϵ. He said he was doing this solely to make his method of calculation intuitively simpler and would eliminate the artificial assumption of discreteness at the end of the calculation, which he did. However, when I read what he had written, it made me gasp with surprise. Here are his words:

> No molecule may have an intermediate or a greater kinetic energy. When two molecules collide, they can change their kinetic energies in many different ways. However, after the collision the kinetic energy of each molecule must always be a multiple of ϵ. I certainly do not need to remark that for the moment we are not concerned with a real physical problem. It would be difficult to imagine an apparatus that could regulate the collisions of two bodies in such a way that their kinetic energies after a collision are always multiples of ϵ.

What Boltzmann could not have imagined is that nature does constantly use an 'apparatus' much like what he suggested; it is quantum mechanics. There's a double irony in this, but I'll save it for Chapter 4; it's better to continue here with Boltzmann's combinatorial argument, which in 1877 he illustrated with just seven particles (which is why I chose seven chequers to illustrate spatial distributions).

For this he supposed the total kinetic energy to be divided up into seven equal portions of magnitude ϵ distributed among the particles. In this case, the chequerboard is replaced by a row of eight bins labelled 0, 1, 2, 3, 4, 5, 6, 7. All particles that have no kinetic energy go into bin 0; those with ϵ (there can be up to seven of them) go into bin 1; particles with 2ϵ go into bin 2 (there can be at most three of them, with one particle in bin 1) and so on. Finally, if one particle has all the energy it goes in bin 7 and the remaining six go in bin 0. Any such distribution corresponds to a macrostate for the kinetic energies. For example, one macrostate could have 3 in the 0-energy bin, 3 in the ϵ bin, and one in the 4ϵ bin. That makes up the specified total of 7ϵ. If, as in the chequer example, each of the particles has its own identity, there are many different ways to put them in the bins. Each choice defines a microstate. Once again, as in the skyscraper analogy, the macrostate defines what the Manhattan skyline looks like and, if bricks with distinct identities are used to build the skyscrapers, the microstate might fix the individual bricks chosen for each.*

In Boltzmann's example with seven particles, there are fifteen different macrostates and 1716 microstates. One macrostate has just 1 microstate—the one in which all particles have the minimal energy ϵ. Another macrostate has 7 microstates; it is the one in which just one particle, for which seven choices can be made, has all the energy and all the remainder have none. One macrostate has 420 microstates; in it, three particles have no energy, two have ϵ, one has 2ϵ, and one has 3ϵ. Two macrostates have 210 microstates each; two have 140; five have 105; three have 42. Thus, just three of the macrostates have 840 out of the total of 1716 microstates: that is only very slightly less than half the total. In fact, the largest macrostate, the one with 420 microstates—its distribution is 3, 2, 1, 1, 0, 0, 0, and it has the greatest number of particles, three, with no energy—already approximates the Maxwell distributions for particles in one dimension (half bell curves) surprisingly well. Only a modest increase in the number of particles would be sufficient to get impressively good agreement.

BOLTZMANN'S BINNED ENERGY distributions illustrate the fundamental role of ratios in physics that I already mentioned in Chapter 1. In Fig. 1 the Maxwellian distributions with different energies look very different but in fact are identical if reexpressed in terms of *relative* energies. They are then all the same as Boltzmann's seven-particle model shows. The bin distributions, the numbers of particles in each bin, are exactly the same whatever the total energy. Measured

---

*One of the most important discoveries that came with quantum mechanics is that elementary particles such as electrons or protons do not have 'personal identities'. Thus, any electron is identical to any other, and the same is true for protons. Moreover, there are two broad kinds of particles: fermions, which include electrons and protons, and bosons, which include the particles of light (photons). The statistical rules that govern the two kinds of particles are different, both from each other and from the rules Boltzmann considered and which apply when quantum effects can be ignored. For the purposes of this book, we only need to consider Boltzmann statistics.

quantities are not absolute but, just like the length of my little fingers, relative.

Let's think a bit more about units; they always derive from something physical. In the metric system, the second derives from the rotation period of the earth; the metre derives from a quarter of the earth's circumference; and the kilogram is defined by a composite platinum-iridium mass held in Paris (every time it is cleaned, it loses a few atoms and, by definition, the universe acquires additional kilogrammes). Thus quantities such as weight, size, and temperature are not intrinsic but always defined relative to something else—a ruler in the case of my fingers. In physics, ratios are the gold standard and always say something about two things.

This means that the concept of thermal equilibrium, which exists in any collection of particles with a Maxwellian distribution, is more fundamental than temperature. One cannot define a temperature for the particles without something to measure it by but one can say they have a thermal distribution. However, if there are two systems, each with such a distribution, it is meaningful to say one has a higher temperature than the other. Everything that has meaning is relative. This unbreakable link will draw us ineluctably to consider our relationship to the whole universe. Carnot sought the most anthropocentrically efficient use of coal, not its ultimate origin. But he did say, "The production of heat alone is not sufficient to give birth to impelling power: it is necessary that there should also be cold; without it, the heat would be useless. And in fact, if we should find around us only bodies as hot as our furnaces, how can we condense steam?" We will see that hot and cold are first simultaneously present in the universe close to the point where time's arrows have their origin.

IN FACT, BOLTZMANN'S macrostates and microstates already start us on the journey to that source of time. Critical for this is the role that probability takes on. We all know that a fair coin will land heads or tails with equal probability and a fair die will come down with equal likelihood on one of its six faces. By analogy, one might suppose that Boltzmann's seven-particle system is a 'die' with fifteen 'faces', one for each macrostate, and that the probability of finding the system in any one of them would be proportional to the number of microstates it contains. Thus the probability of finding the largest macrostate is 420/1716 and the smallest is just 1/1716. Both of these probabilities as well as those for the other thirteen macrostates are well defined because the total number of microstates is finite. Boltzmann's great innovation was to define entropy in terms of probability. The number of microstates that a particular macrostate has will determine its probability of occurrence but not which of the microstates is actually realised. This makes entropy a measure of ignorance. It is also important that probability arguments presuppose that the total number of possibilities is finite. You would not want to bet on the fall of a coin with infinitely many sides.

To get from probabilities to the notion of entropy that developed out of Boltzmann's work, we need logarithms, those devices that so usefully reduced multiplication to addition when John Napier, a Scot, introduced them in 1614. Logarithms, abbreviated 'log', must have a base. If it is 2, as in the binary system, then the log of 32 is 5 because $2^5 = 32$. The log of 4 is 2 because $4 = 2^2$; the log of 8 is 3 because $2^3 = 8$. Then to multiply 4 by 8 you simply add 2 to 3, getting 5, which is the log of 32. In the decimal system, the base of logarithms is 10 and the log of 100 is 2, of 1000 it is 3, and so on.

The great value of logarithms for defining entropy is the way they can be added if one is considering two or more systems that are independent, so that there are no statistical correlations between them. The reason for this is that the probability of independent events occurring simultaneously is the product of the individual events. Thus, the probability of one coin landing heads is 1/2 and of another is also 1/2. But the probability that if tossed simultaneously they both land heads is $1/2 \times 1/2 = 1/4$. In thermodynamics, the entropy of a system formed by combining two previously independent systems is the sum of their entropies. This can be matched in statistical mechanics if the entropy of a system is made to be proportional, not directly to its probability, but to the logarithm of its probability.

The addition of logarithms also applies to a single dynamical system of particles if there is no correlation between their positions and momenta. In the two illustrative examples considered previously, the first is for the configuration space and the second for the momentum space. For the former all particle positions were assumed to be on an equal footing, which is justified only in the case of an ideal gas, for which potential energy is not taken into account and is assumed to be zero. This leads to flat (uniform) spatial distributions as the most probable. The possible kinetic energy distributions are different because the total energy is fixed; this skews the possible distributions into the Maxwellian form of Fig. 1 and the non-uniform binned distributions of Boltzmann's example. In the case of an ideal gas, there are no correlations between the particle positions and their momenta, and in this case the logarithms of the respective probabilities can be simply added to find the total entropy.

The German word for 'probability' is *Wahrscheinlichkeit*. In Vienna's Central Cemetery, Boltzmann's tombstone bears a bust of him and the equation for entropy, $S = k \log W$, which he actually never wrote, chiselled above it. Have a look online; along with $E = mc^2$, the Boltzmann relation is rated one of the most important in physics. The parallel justifies Clausius's coining of 'entropy' to make it sound like 'energy'. Entropy reflects the extent to which energy is divided among *dynamical degrees of freedom*, Maxwell's generic term to cover all forms of mechanical energy whether in particulate or wavelike form. The k, which is called the Boltzmann constant, could be omitted; it comes in when units are employed to measure energy. As in the seven-particle macrostates, entropy can be reduced conceptually to simple counting of microstates, which in fact is literally true in quantum mechanics.

WE ARE NEARLY at the end of this part of the story. It can be summed up as the microscopic explanation of

$$S = \oint \frac{\mathrm{d}Q}{T} \quad \text{by} \quad S = k \log W.$$

It is often said that all truly great ideas in physics are simple. I hope I have explained Boltzmann's well enough for you to agree. However, it remains to describe more fully Boltzmann's insight (which he expressed rather tersely) into what happens in equilibration and its concomitant growth of entropy. This is where the connection between statistics and dynamics is established. I expect many of my readers will have read one of Roger Penrose's books, beginning with *The Emperor's New Mind*. Penrose uses beautiful diagrams to divide up phase space pictorially, on the two-dimensional page, into different possible macrostates with different numbers of microstates. It would be impossible to make their areas proportional to the numbers of microstates, since these differ hugely. However, logarithms do help; if 10 is their base, a macrostate with 1000 times the number of microstates than another is only 3 times as large. Even so, Penrose warns his readers that the differences are still huge, especially for the thermal equilibrium macrostate. Even with logarithms, it is a lot larger than all the others put together.

This fact is the key to the statistical explanation of equilibration. Suppose that, for some unexplained reason, a dynamical system is, at some initial time, in a highly improbable (very 'small') macrostate. So far I have described the macro- and microstates at only a given instant. It's now time to introduce the notion of a *microhistory*. This is the path traced out by the system in its phase space as time passes. The actual path the system takes depends with extreme sensitivity on the individual interactions of its particles with each other. Through macroscopic observations one cannot possibly determine the microscopic state with precision remotely adequate to predict the course of any microhistory. One can do that for the planets of the solar system but not for 10^{23} molecules in a gas. For the gas, one can only use probabilistic arguments to surmise what is likely to happen in a microhistory.

Now, phase space is not like our familiar space in which you can go in only three directions. For a gas of 10^{23} molecules its phase space has 6×10^{23} dimensions, and the system, inexorably driven forwards by the dynamical law that governs it and buffeted by all the collisions between the molecules, can go in as many directions, constantly swapping from one to another. It takes little imagination to realise that if the system at some instant is in a small macrostate it is unlikely to wander from it into a smaller one; it is much more likely to go into a larger one with greater entropy and, after what is called its *relaxation time*, to be in the macrostate of thermal equilibrium. From there it has virtually no chance of escaping. It was through an insight like this that Boltzmann brought probability into dynamics despite the fact that it is a strictly deterministic science.

He also noted a curious corollary of his explanation. Suppose that at the initial instant, when the system is in the small macrostate, we suppose all the velocities exactly reversed but the positions unchanged. This is the standard transformation of time-reversal symmetry. If we then work out what the system's microhistory will do, we shall learn not what happens in its future but what happened *in its past*. Of course, the system cannot know whether it is going backwards or forwards in time. It simply does what the dynamical law which governs it dictates. It will therefore, with overwhelming probability, find its way in exactly the same manner as when going forwards in time, albeit not through the same macrostates. Its entropy will increase both backwards and forwards in time from the special initial low point. Penrose explains this very well. In fact, even without Boltzmann's clear insight into the statistical predominance of uniform and thus high entropy states, Thomson had already noted in 1874 that such would be the behaviour of molecules in a box. Boltzmann went further and said the universe would behave in the same way. That claim brings with it lots of issues and takes us to Chapter 4. I will end this one by noting that if Hamlet's nutshell really were an infinite space nothing could equilibrate in it.

CHAPTER 4

BOLTZMANN'S TUSSLE WITH ZERMELO

The Background

This chapter describes the conclusion of Boltzmann's work on the second law. It sets the scene for the rest of the book by showing why, despite his brilliant ideas, Boltzmann did not succeed in finding a fully satisfactory resolution of the conflict between the time-reversal symmetry of the laws of nature and the manifest temporal asymmetry of all the macroscopic phenomena we observe around us. All this became clear in a famous debate in the 1890s between him and the young Ernst Zermelo, who later became a distinguished logician and made important contributions to the foundations of mathematics.

It will not go amiss to put the exchange between Boltzmann and Zermelo in its proper perspective. Some famous debates have marked the history of physics. The first was Galileo's argument with the Catholic Church—and many sceptics—about the truth of Copernicus's proposal that the earth orbits the sun and not the sun the earth. A century later came the debate about the nature of time and motion: are they absolute, as Newton argued, or relative, as Leibniz insisted? More recently there was the debate between

Einstein and Niels Bohr about the interpretation of quantum mechanics. The only other debate of similar significance that comes to mind is the one that will be the subject of this chapter. The argument about the relative merits of geocentricity as opposed to heliocentricity has long been settled in favour of Copernicus and Galileo, but the other three—all interrelated, I will argue—still await resolution.

In the decade and a half that followed Boltzmann's 1877 insight, majority opinion among German physicists turned, rather surprisingly, against atomism and the statistical-mechanical interpretation of thermodynamics. One reason was the failure of theory, despite many successes, to match experiment fully. This applied especially to the problem that a given amount of heat added to molecular gases confined in a fixed volume was found to raise its temperature more than expected. Temperature was known to reflect mechanical activity, which theory predicted should take the form of not just the translational and rotational motion of molecules but also their internal vibrations. The effect of these last did not show up. Even more serious was the miserable failure of theory to explain the spectral lines of the radiation emitted and absorbed by atoms and molecules. What nobody realised was that models based on classical mechanics had no chance of resolving these problems. That had to wait until the full discovery of quantum mechanics in the mid-1920s by Heisenberg and Schrödinger. Thus, failures fully understandable with hindsight were fostering a certain distrust in atomism.

Perhaps more important was the philosophical movement known as positivism. Several prominent thinkers on the continent, especially in Germany and Austria, favoured it. The best-known among them was Ernst Mach, whom I quoted in Chapter 1. He had a strong distrust of theory because his study of the history of science had shown how many seemingly plausible theoretical models proved under testing to be inadequate to describe phenomena. Caloric is a good example. Like other positivists, Mach argued that the only proper task of science was to identify and describe phenomena which one could repeatedly observe and describe by means of measurement. Distrust in atomism peaked in the 1890s, not only for the two reasons already given but also on account of doubts about the ability of mechanical models to explain irreversible phenomena, above all the growth of entropy.

Although he was careful not to prescribe precise models of atoms, Boltzmann had for decades been a leading advocate of atomism and was challenged for this reason. At the end of a lecture Boltzmann gave at the Imperial Academy of Science in Vienna in 1897, Mach notoriously said loudly from the audience, "I don't believe that atoms exist". This was bad for Mach's posthumous reputation, though it may be said in his defence that when the existence of atoms was definitively established they turned out to be quite unlike what had been imagined. Indeed, their actual nature, or rather how they connect with the rest of the universe, is still subject to the mysteries that to this day shroud quantum mechanics. As for the unfortunate Boltzmann, it is argued that criticism like Mach's contributed to his desperately sad suicide in 1906. However, his well-attested mood swings suggest he was bipolar, and that may have been the main reason he did not live to experience the well-deserved widespread recognition of his work that began just at the time of his death.

In fact, although in the 1890s he clearly felt he was a prophet unrecognised in his own German-speaking lands, he was taken much more seriously in Britain and in 1894 was invited to attend a meeting of the British Association in Oxford, at which there was a serious discussion of his *H*-theorem and its apparent conflict with time-reversal symmetry that Loschmidt had raised. This meeting stimulated Boltzmann to write a letter to *Nature* (in remarkably good English) in which he addressed various questions in the philosophy of science and concluded with a passage that anticipated his last stand defending the virtues of atomism and the mechanical explanation of entropy.

Commenting on the criticism that the *H* theorem could not be true, for if it were "every atom of the universe would have the same average *vis viva*, and all energy would be dissipated", Boltzmann countered that "this argument only tends to confirm my theorem, which requires only that in the course of time the universe must tend" to such a state. Thus there was nothing wrong with the *H*-theorem; the great mystery was "why this state is not yet reached". He concluded his letter with an idea he attributed to his old assistant, a certain Dr Ignaz Schuetz:

> We assume that the whole universe is, and rests for ever, in thermal equilibrium. The probability that one (only one) part of the universe is in a certain state, is the smaller the further this state is from thermal equilibrium; but this probability is greater, the greater is the universe itself. If we assume the universe great enough, we can make the probability of one relatively small part being in any given state (however far from the state of thermal equilibrium), as great as we please. Assuming the universe great enough, the probability that such a small part of it as our world should be in its present state, is no longer small.
>
> If this assumption were correct, our world would return more and more to thermal equilibrium; but because the whole universe is so great, it might be probable that at some future time some other world might deviate as far from thermal equilibrium as our world does at present. Then the aforementioned *H*-curve would form a representation of what takes place in the universe. The summits of the curve would represent the worlds where visible [macroscopic] motion and life exist.

Schuetz's idea is a remarkable—and possibly the first—formulation of what is now known as the *anthropic principle*, which plays an important but hotly contested role in modern physics and cosmology. It is invoked to explain why the laws of physics and the state of the universe appear to be so finely tuned that intelligent life is possible. The principle simply says that if they were not we would not be here to register the fact. This is the clear implication of the above passage. We will see that Boltzmann made some significant additions, with far-reaching implications for the very nature of time, to Schuetz's suggestion a couple of years later.

Sandals in a Sandpit

The rest of this chapter and much of the book from now on deal with the implications of two theorems. I won't attempt to give their proofs, only an idea of what is involved. I will have succeeded if at the end of my account you understand the connection with sandals. To get you thinking in the right way, imagine a girl in sandals who starts walking in a freshly raked sandpit and seeks never to step where the sandal on, say, her right foot has already left a print. Because the sandpit, like the sandals, has a finite size, there must come a time when the girl can no longer find a place where she has not already trod with her right foot. She could always avoid stepping on her original footprint, but to avoid places where she has already stepped, she must step in some new place. After a while, there will simply be no more places to step. This will be so no matter how small the size of the sandals relative to the size of the sandpit.

It was a fact as simple as this which gave Boltzmann great difficulties in the 1890s; they are still unresolved. The sandpit stands for phase space, the key concept I asked you to keep in mind. For the molecules of gas studied in any laboratory experiment, it has that immense number of dimensions, of order 10^{23}, and is virtually impossible to visualise. The microhistory of the dynamical system under consideration is traced out by a *representative point*, the instantaneous phase, that 'wriggles' its way through phase space. Now suppose that a pin stuck through the bottom of the girl's sandal is taken to mark one of the successive representative points in the sandpit—the phase space—of such a microhistory. Because we can suppose the sandals to be as small as we like provided their size is greater than zero, this means that as the representative point traces out the microhistory, it cannot avoid returning at some stage arbitrarily close to any point it previously visited. Indeed, it cannot avoid doing so infinitely many times.

Next I must say something about the size of the sandals. In statistical mechanics there is a concept known as a Gibbs ensemble (it was introduced by Willard Gibbs, who quoted Clausius's two laws of the universe). To define such an ensemble, one imagines many identical systems—that is, systems all governed by exactly the same dynamical law—whose states at a given instant of time correspond to points in phase space that belong to a connected region of phase space. Now, the dynamical laws employed in statistical mechanics and assumed for the purposes of the present discussion have a remarkable consequence when applied to a Gibbs ensemble. Although the shape of the ensemble can change, its volume—the sandal size in the metaphor—must remain exactly constant. This is the consequence of a result proved by Joseph Liouville, though in fact Boltzmann was, in the two or three years before he developed his *H* theorem, the first theoretician to make significant use of the result and caused it to be known as Liouville's theorem. Willard Gibbs, in a book published in 1902, just before he died, made it into the central unifying theorem that underlies statistical mechanics. To get a better idea of what Liouville's theorem implies, you can liken a Gibbs ensemble to a patch of oil on the surface of a stream. As time passes, the patch is carried down the stream without breaking up; though its shape can change, its area must remain constant.

Liouville's theorem holds that whatever the number of dimensions that phase space has, it must be an even number, since for each degree of freedom corresponding to position there is a matching one for momentum. Importantly, the theorem holds for what are called Hamiltonian dynamical systems, which occur widely in nature under conditions for which frictional effects can be ignored, as is the case with the (idealised because frictionless) pendulum I used to illustrate the ebb and flow of energy conservation; Hamiltonian systems were and still are widely used to model the behaviour of particles in statistical mechanics.

By now you should begin to see the connection between the theorem and sandals in the sandpit.* The analogy is limited but, I think, helpful. The sandpit stands for phase space and the fixed sandal size for the dynamically enforced constant size of a Gibbs

*I am indebted to Mr Jim Wheeler for telling me about it in an email exchange.

ensemble. If phase space has a finite size—the technical expression is that it has a *bounded measure*—the ensemble cannot avoid coming back again and again to any region it previously visited. That's more or less what happens to the girl in the sandpit. In the general multidimensional case the ensemble can, like an octopus, 'grow tentacles', and the simple sandpit analogy with fixed sandal size breaks down. The great French mathematician Henri Poincaré, who himself harboured strong doubts about atomism, proved a sophisticated generalisation with his famous *recurrence theorem* in 1891. He considered an ensemble that initially occupies only a very small volume of phase space. Liouville's theorem is assumed to hold and, critically, phase space must have a bounded measure.

With these two assumptions he proved a truly remarkable result: however small the initial patch may be, any point within it that is the representative point of the actual microhistory of the system being considered and is being carried along by the flow in phase space will (except for some contrived cases that can be ignored because they are so unnatural) return infinitely often arbitrarily close to the initial point in phase space provided only that one waits long enough. In statistical mechanics with anything more than a very few particles the *recurrence times* very soon become immensely long (much greater than the age of the universe), and they become infinite as the number of particles tends to infinity. Just as in the sandpit case, the result of the theorem holds because the volume of the patch, however small it may be, eventually runs out of phase-space volume that it can 'visit'. As we will now see, it is this theorem that gave Boltzmann such difficulty in his debate with Zermelo.

Zermelo's Objections and Boltzmann's Responses

When he initiated the debate, Zermelo was a student of Max Planck, a distinguished professor in Berlin and a considerable expert on thermodynamics; among other things, he introduced a useful sharpening in the formulation of the second law. He was a cautious, somewhat conservative theoretician who preferred purely empirical phenomenological thermodynamics to attempts at an atomistic interpretation of its results. This is why he encouraged Zermelo to submit a paper criticising Boltzmann's theories to the *Annalen der Physik*. Before I describe the ensuing debate, I will explain the double irony I promised in connection with Boltzmann's 'anticipation' of quantum mechanics when excusing his introduction of discrete amounts of kinetic energy to count microstates.

By the middle of the 1890s a crisis had developed in the theory of radiation. Application of Maxwell's theory of the electromagnetic field had led to a manifestly nonsensical result. Radiation in thermal equilibrium in an enclosed cavity should have an infinite amount of energy, but if that were the case it would be impossible to heat an oven in order to bake a loaf of bread! This was manifestly nonsense. At the same time possible technical applications meant that much interest attached to the exact spectral distribution of the radiation when in equilibrium; it is called *blackbody radiation* because it has the properties of a body that at every wavelength absorbs all radiation, reflecting none. From the mid-1890s Planck started to try to determine theoretically the form of the blackbody spectrum of radiation in a container using known phenomenological relations, taking the second law of thermodynamics to be fundamental, like the first law. He failed. This forced him to try Boltzmann's statistical methods, including partial adoption of energy quantisation, or rather, considering amounts of radiation in energy bins. This worked, and Planck published his result in 1900. It marked the birth of quantum mechanics, perhaps the greatest of all revolutions in physics.* For an introduction to Planck's work and references to fuller accounts, see Jacobson's paper cited in the bibliography.

Now for what Zermelo argued in his first paper, which appeared in 1896 and began with a proof of the recurrence theorem, which

---

* Compounding his misfortunes, Boltzmann does not seem to have followed developments closely starting around 1900. They included one of Einstein's famous papers of 1905, which greatly strengthened the evidence for molecules and atoms and in fact used a possibility—the effect of fluctuations—that Boltzmann had mentioned in his debate with Zermelo. Perhaps because he felt guilty about his role in Zermelo's criticism of Boltzmann, Planck, defending the existence of atoms in a vigorous volte-face, mounted a furious attack on Mach in 1909.

he showed must hold "provided that the coordinates and velocities cannot increase to infinity". In fact he did little more than raise the same problems to which Poincaré had already drawn attention in papers both before and after publication of the recurrence theorem. As all practitioners of statistical mechanics had always done, Zermelo assumed gases whose molecules were enclosed in a container with rigid walls off which they bounced elastically. In such a situation, the conditions of the recurrence theorem are satisfied and therefore the gases, "instead of undergoing irreversible changes, will come back periodically to their initial states as closely as one likes". Since quantities like the temperature and entropy are determined in accordance with the mechanical theory by the instantaneous state, irreversible changes of them cannot be expected to occur.

Boltzmann had not responded to Poincaré's papers, but Zermelo's did rouse him. He granted that Poincaré's theorem was clearly correct but that Zermelo's application of it to the theory of heat was not. Moreover, Boltzmann had always "emphasised that the second law of thermodynamics is from the molecular viewpoint merely a statistical law. Zermelo's paper shows that my writings have been misunderstood; nevertheless, it pleases me for it seems to be the first indication that these writings have been paid any attention in Germany".

The gist of Boltzmann's response is simple and convincing. Expressed in terms of entropy—Boltzmann actually discussed the equivalent behaviour of his *H*-function—it is that in a container at rest holding a great number of molecules their state will for vast stretches of time be very close to the most probable one of maximal entropy. There will always be tiny, essentially unobservable deviations from it, but just occasionally there will be large ones. The time between the occurrences of such large deviations will be many, many times greater than their duration. Necessarily, then, it is true that the entropy can decrease and that therefore the molecular theory of heat cannot derive the second law as an absolute law valid under all circumstances. Indeed, for a small number of the many molecules in that container (and keep in mind that molecules had never yet been observed) violations must be expected to occur frequently. (It was the existence of the associated fluctuations that Einstein exploited in 1905.)

However, the real question was whether the mechanical viewpoint led to any contradiction with experience—for example, a system being seen to pass from a state of low entropy to one of high entropy and then return to the original low-entropy state in an observable length of time. If not, we would never actually see any such violation of the second law. Boltzmann gave a simple example involving "a trillion tiny spheres" that had a recurrence time long enough to make "any attempt to observe it ridiculous". Zermelo, surely egged on by Planck, was a true gadfly, and that was not the end of their argument. But before we come to the second attack, two further comments in Boltzmann's first response are important.

In the first, turning briefly his attention from molecules in a box, Boltzmann said: "Naturally, we cannot expect from natural science an answer to the question—how does it happen that at present the bodies surrounding us are in a very improbable state—any more than we can expect from it an answer to the question why phenomena exist at all and unfold in accordance with certain given laws". Although, like Boltzmann, I suspect that any answer to his second question, if it exists at all, is far beyond our present ken, I do think there is a satisfactory answer to the first. It is why I have written this book. In fact, despite his pessimism, I think that in his debate with Zermelo, Boltzmann already gave some hints about where the answer to his first question, and with it explanation of time's arrows and the second law, is to be sought: through consideration of the universe in its totality and not just things we can observe around us. It is not that he says much that is explicit. But, having noted Poincaré's theorem implies the entire universe must return to its original state after a sufficiently long time, he asks: "How shall we decide, when we leave the domain of the observable, whether the age of the universe, or the number of centres of force which it contains is infinite? Moreover, in this case the assumption that the space available for motion, and the total energy, are finite, is questionable".

Brief as these comments are, they are critical, especially about the space available for motion. If it is not finite, the phase space of the universe does not have a bounded measure and the recurrence theorem fails. The system we are trying to understand is no longer in a box. The girl walks in an infinite sandpit. The implications are considered in the following chapters; this one continues with the 'battle of the second law' as it played out in the pages of the *Annalen der Physik*.

Zermelo's response to Boltzmann's response came soon. He granted he had not been fully familiar with "Herr Boltzmann's investigations of gas theory" and agreed that one could choose either the principle that entropy never decreases or the fundamental modification entailed in the mechanical viewpoint. But "as for me (and I am not alone in this opinion), I believe that a single principle summarising an abundance of established experimental facts is far more reliable than a mathematical theorem, which by its nature represents only a theory which can never be directly verified; I prefer to give up the theorem rather than the principle, if the two are inconsistent".

A little later, Zermelo raises, rather effectively, a new issue. He notes that the discussion concerns "the entropy of *any arbitrary* system free of external influences. How does it happen, then, that in such a system there always occurs only an *increase* of entropy and *equalisation* of temperature and concentration differences, but never the reverse? Probability theory cannot help here, since every increase corresponds to a later decrease, and both must be equally probable or at least have probabilities of the same order of magnitude". It follows that the initial state one would find on examining a system of molecules subject to Poincaré recurrence could just as well lie in a time interval in which the entropy is decreasing as in one where it is increasing. How then does it come about that one only always observes an increase of entropy? He argues that "as long as one cannot make comprehensible the *physical origin* of the initial state, one must merely assume what one wants to prove". Zermelo hit the mark with this observation. It bothers people to this day. A closely related problem, which I think Zermelo also had in mind when invoking any arbitrary system, concerns two independent spatially separated systems: in one the entropy could be increasing, in the other it could be decreasing. That too was never observed. The harsh implication was that statistical mechanics could not give a satisfactory explanation of the universally observed growth of entropy.

Predictably, Boltzmann's response to Zermelo's response followed in the next volume of *Annalen der Physik*. I will come in a moment to the subtle argument by which Boltzmann countered Zermelo's argument that entropy decrease is as likely as entropy increase. But I first want to give what can be thought of as the two options Boltzmann saw at the end of his life as the best ultimate mechanical explanation of the second law of thermodynamics. To this day, almost all such explanations that have been considered boil down to one or the other of his two proposals.

The first came right at the start of the paper, where he suggested that "the universe—or at least a very large part of it which surrounds us—started from a very improbable state, and is still in an improbable state. Hence, if one takes a smaller system of bodies in the state in which he actually finds them, and suddenly isolates this system from the rest of the world, then the system will initially be in an improbable state, and as long as the system remains isolated it will always proceed toward more probable states". Without the restriction to "a very large part" (which does not help), this first proposal of Boltzmann's—namely, that the universe began in a very improbable state—is identical to the past hypothesis. As explained in Chapter 1, the past hypothesis involves a more or less ad hoc addition to the known laws of nature that requires the universe to have started its existence with an initial condition much more special than anything warranted by the form of the laws. This is not satisfying. It means that within the structure of the laws of nature, no trace can be found of an explanation for what is perhaps the most striking fact of existence—the profound difference between the past and the future. Nevertheless, Boltzmann's special initial condition has been taken seriously now for several decades and is favoured compared with his second proposal. It is essentially what he had attributed in *Nature* to Schuetz but with amplifications sufficiently important to warrant full quotation:

> In a sufficiently large universe which is in thermal equilibrium as a whole and therefore dead, there must be here and there relatively small regions of the size of our galaxy (which we call worlds), which during the relatively short time of eons deviate significantly from thermal equilibrium. Among these worlds the state probability increases as often as it decreases. For the universe as a whole the two directions of time are indistinguishable, just as in space there is no up or down. However, just as at a certain place on the earth's surface we can call 'down' the direction toward the centre of the earth, so a living being that finds itself in such a world at a certain period of time can define the direction of time as going from the less probable to more probable states (the former will be the 'past' and the latter the 'future') and by virtue of this definition he will find that this small region, isolated from the rest of the universe, is 'initially' always in an improbable state. This viewpoint seems to me to be the only way in which one can understand the validity of the second law and the heat death of each individual world without invoking an unidirectional change of the entire universe from a definite initial state to a final state.

The "only way" in the last sentence clearly indicates Boltzmann's preference for a resolution of the problem of time's arrows that respects the time-reversal symmetry of the laws of nature as opposed to a time-asymmetric special initial condition. Indeed, in the second volume of his *Lectures on Gas Theory*, published two years later, he repeats the preceding passage more or less verbatim, prefacing it with the statement that anyone who tries to explain "the apparent unidirectionality of time" must use "equations in which the positive and negative directions of time are equivalent". Although it was Boltzmann who was the first to moot tentatively what is now called the past hypothesis, it was clearly not his preferred explanation of time's unidirectionality.

It was through the 'tweaks', in just two or three brief comments in the quoted passage and his later book, that Boltzmann persuaded scientists that the experienced direction of time is aligned with the direction of entropy increase. The 1895 letter to *Nature* did not include the argument that the less probable state will be taken to be the 'past'. This is one of the ideas for which Boltzmann is famous; it seems to have been his own addition to Schuetz's original idea. Reading Boltzmann's 1877 paper and Thomson's of 1874, both of which had bidirectional directions of entropy increase from a minimum, I do not get the impression that either scientist, in that decade, was capable of shedding the instinctive feeling that, whatever may be happening in a box or the universe at large, time flows forwards inexorably.

Also, I am not aware that Boltzmann ever made it fully explicit in the 1890s that intelligent beings could exist on both sides of a single entropy dip and therefore be living in a spatially and temporally localised region of the universe with bidirectional arrows of time. However, near the end of the section 'Application to the universe' in the second part of his *Lectures on Gas Theory*, having repeated some of the comments made in the exchange with Zermelo, he does say that if unidirectional change of the entire universe from a definite initial to a final state does not occur the situation will be as follows:

> In the entire universe, the aggregate of all individual worlds, there will however in fact occur processes going in the opposite direction. But the beings who observe such processes will simply reckon time as proceeding from the less probable to the more probable states, and it will never be discovered whether they reckon time differently from us, since they are separated from us by eons of time and spatial distances $10^{10\wedge 10\wedge 10}$ times the distance to Sirius—and moreover their language has no relation to ours.

This is very close to saying explicitly that there will be bidirectional arrows associated with a single entropy dip.

If the part of Boltzmann's response just considered, which effectively destroyed Newton's notion of absolute time that flows uniformly "without relation to anything external", has been widely accepted and is seen as a significant advance in the understanding of time, it is not possible to say the same about his answer to Zermelo's killer argument: why don't we see entropy decrease around us as often as we see it increase? In fact, Boltzmann made no direct attempt to counter the argument. He merely said that large deviations from maximal entropy are enormously rarer than small ones. What he might have said but didn't is that if you do manage to catch a state with low entropy it will very soon, if not immediately, return to one of maximal entropy even if initially the entropy does decrease. This is because departures from maximal entropy get vastly more improbable the greater the departure and, as Boltzmann had already argued in 1877 (and as Penrose has made so plausible with his diagrams), such a system has a built-in tendency, for both directions of its velocities, to find its way to equilibrium rapidly from any out-of-equilibrium state.

Boltzmann's failure to provide a convincing answer to Zermelo's most serious charge brought their exchange to an inconclusive end. It had, however, clarified some important issues and introduced new ideas, especially the promising one that it is increase of entropy which determines the experienced direction of time. Borrowing terms from music, one can say that the debate brings to a close the exposition of the problem. Before I start on my proposed resolution of it, we need a brief recapitulation at the end of this chapter to highlight certain aspects and also, in Chapter 5, a transition to illustrate the remarkable persistence of the 'steam-engine paradigm' right into the twenty-first century.

The Outstanding Issues

It is my belief that the recurrence theorem has led many people, beginning with Zermelo and Boltzmann, to think about entropy increase in quite the wrong way (as I already suggested in Chapter

1). We need to consider the conditions under which the problem is formulated. The situation with particles in a box with impenetrable walls, for which Poincaré's theorem does hold, is very artificial even though it proved to be an idealisation that was fruitful almost beyond measure. One only has to mention Boltzmann's beautiful count-of-microstates explanation of Clausius's phenomenological $dS = dQ/T$ definition of entropy increments, the strong evidence that atoms and molecules with definite sizes exist, and the discovery of quantum mechanics to be reminded of that.

The problem lay elsewhere. Boltzmann identified it when he said science could not explain why "at present the bodies surrounding us are in a very improbable state". Zermelo echoed this when he said no progress could be made until one can "make comprehensible the *physical origin* of the initial state". That we can create in the laboratory such initial states, which are far from being in thermal equilibrium with their environment and therefore are all in a very improbable state, is solely thanks to what I call the 'gifts of nature', by which I mean the solid objects out of which we can make boxes, rulers to measure their sizes reliably, and clocks to time experiments. These things too are far from thermal equilibrium with their environment and so in a very improbable state. Where do these gifts of nature come from? The answer to that question is an integral part of the problem.

Boltzmann was entirely justified in saying that the mechanical theory of heat in conjunction with probability arguments provided a fully adequate explanation of equilibration. The problem is not that we never see an equilibrated system disequilibrate; it is the lack of equilibrium all around us that is the problem. Carnot said, "The phenomenon of the production of motion by heat has not been considered from a sufficiently general point of view". The problem we face now is not the behaviour or nature of heat; that has long been clarified. What we need is a point of view that enables us to understand where all the bodies in highly improbable states around us come from.

Since the 1960s it has been suspected that this will require us to shift our view from the laboratory to the universe. As we have just seen, Boltzmann's writings already contain hints in that direction, but there was hardly any way in which useful ideas about the universe could have been obtained in the nineteenth century; the necessary theory and above all observations were completely lacking. However, despite extraordinary advances in cosmology since the expansion of the universe was discovered nearly a century ago, when it comes to the origin of the entropic arrow no suggestions that go significantly beyond the two that Boltzmann advanced to Zermelo have been made since their debate. It is true that the expansion of the universe has, since a paper in 1962 by Thomas Gold (the joint creator with Hermann Bondi and Fred Hoyle of the steady-state theory of the universe), been conjectured to be the 'master arrow' behind all time's arrows, but I am not aware of specific mechanisms that do not, in one way or another, invoke the ideas that go back to Boltzmann.

What I find particularly surprising about the post-Boltzmann study of time's arrows has been the apparently unconscious transfer of arguments appropriate under conditions when the recurrence theorem will hold to circumstances for which that seems very questionable. This is perhaps the most striking manifestation of the 'steam-engine paradigm', to which I devote Chapter 5. It will have the added advantage of showing what happened to the fluctuation idea first suggested by Schuetz before being given wide currency by Boltzmann. It still plays a surprisingly large role in modern cosmology.

CHAPTER 5

THE CURIOUS HISTORY OF BOLTZMANN BRAINS

BOLTZMANN BRAINS ARE BIZARRE SPECTRES THAT HAUNT THE MINDS OF modern cosmologists who try to make secure predictions about what they observe in the universe if, as several theories suggest, it is only part of a multiverse that is infinitely large. This has an implication even more troubling than eternal recurrence. The nightmare in the most extreme case envisaged is that exactly what you firmly believe yourself to be doing now, in this very instant and in every last detail, is being replicated infinitely many times by infinitely many identical avatars of yourself scattered around the infinite multiverse, every one of them believing themselves to be the one and only true you. What is more, each of them, including almost certainly you, consists of nothing more than a disembodied brain that exists fleetingly in what is otherwise fluctuating chaos. This matters for cosmologists who are trying to explain why we take it to be fact that we are not disembodied brains but live in a highly ordered universe with a history which stretches back 13.7 billion years to a big bang. Science is meant to explain phenomena. How can it have gone so spectacularly wrong? As we will see in this chapter, the issue is intimately related to a serious difficulty that came to light in the 1930s with the Schuetz-Boltzmann fluctuation idea and has since created problems in modern cosmology.

In brief, what happened is that in the early 1980s a theory called inflation (to which I will come in Chapter 18) achieved spectacular success in explaining the large-scale structure of the universe—in particular, how galaxies and clusters of galaxies are distributed—but at the price of increasing the size of the universe vastly beyond what had previously been imagined. What is more, the theory of inflation itself suggested that if it could work once it could go on happening again and again, creating one universe after another within an infinite multiverse. The question then arose of how such a theory could be tested. This involved identifying what is called a 'typical observer'. Theorists then recognised the uncomfortable possibility that such an observer might not be a right-thinking astronomer in an actual universe but simply a disembodied brain that happened to fluctuate into and then out of existence. Around the turn of the present century, a connection was recognised between this acute dilemma and the Schuetz-Boltzmann fluctuation idea.

It's not my aim in this chapter to discuss the merits or demerits of the multiverse idea. I am only interested in the history of the original fluctuation idea from when it was floated in *Nature* in 1895 to the coining of the expression 'Boltzmann brain' in 2004. I do this for two reasons. The first is to demonstrate the persistence of the 'steam-engine paradigm' long after one might think it should have been abandoned or at least seriously questioned. The second is to prepare the ground for a tentative suggestion related to the situation (mentioned in Chapter 1) in which the size of the universe at the Janus point is zero. If the suggestion is correct, it has potentially spectacular implications that I will present near the end of the book.

A first curiosity about the original fluctuation idea is its premise: what grounds did either Schuetz or Boltzmann have for assuming that "the whole universe is, and rests for ever, in thermal equilibrium"? The truth is that not only now but also in Boltzmann's time the conditions on the earth and all around it out to the most distant stars are, and were then, extraordinarily far from thermal equilibrium. Moreover, the equilibrium states we observe around us have invariably come about through equilibration. Doesn't that point to an earlier state of extreme disequilibrium that has not yet had sufficient time to equilibrate? Boltzmann did indeed make a tentative suggestion along those lines, as we saw in Chapter 4. It seems that his desire, also noted there, to have a time-symmetric universe was a significant factor in his retention of the fluctuation hypothesis.

Two other factors could explain why Boltzmann felt the assumption was reasonable. First, despite growing evidence for geological history and increasing acceptance of Darwinian evolution of species, a belief (perhaps unconscious) in an overall unchanging universe was widespread. Even Einstein chose a stationary model of the universe when he laid the foundations of relativistic cosmology in 1917, and as late as the mid-1920s he roundly rejected expansion of the universe even though his own theory of general relativity clearly allowed such solutions, as he well knew. Second, a great proportion of Boltzmann's life's work had been devoted to understanding the nature of thermal equilibrium. Under the conditions in which it had been mostly investigated—an ideal gas confined by the walls of a box off which the molecules bounce elastically—he had argued persuasively that (to phrase it in modern terminology) an initially out-of-equilibrium macrostate of the gas would with overwhelming probability evolve in either time direction to the macrostate of thermal equilibrium. Penrose's diagrammatic argument for this conclusion in modern terms is very powerful.

I think it is therefore reasonable to assume, given the total absence in the nineteenth century of information about the universe on the very largest scales, that both Schuetz and Boltzmann were unconsciously transferring concepts and methods well justified under laboratory conditions—they largely survived the quantum revolution and were critical in the discovery by Planck that unleashed it—into a realm about which they could make only the vaguest of guesses. Let us now consider what happened to the fluctuation idea.

THE FIRST THEORETICIAN worth discussing in this connection is Eddington, who made that blunt warning to any theoretician foolhardy enough to challenge the second law of thermodynamics. He also said, "Shuffling is the only thing which Nature cannot undo". In a supplement to *Nature* in 1931 (based on an address he had given to the Mathematical Association) he considered—though without mentioning Boltzmann—a possible loophole that would reconcile entropy increase with time-reversal symmetry. He said: "If we have a number of particles moving about at random, they will in the course of time go through every possible configuration, so that even the most orderly, the most non-chance configuration, will occur by chance if we wait long enough". Eddington is here assuming a condition stronger than the recurrence theorem (that *previously realised* states will be revisited arbitrarily closely infinitely often)—namely, the ergodic hypothesis that *all* states will be visited.

Now, it is very easy to imagine situations in which Eddington's claim is manifestly false. The simplest example is one in which a very large number of unconfined particles move purely inertially. As I will show in Chapter 6, it is easy to prove that in such a situation there is a unique instant at which the overall size of the system of particles passes through a minimum value and in both directions of Newtonian time grows to infinity while the distances between the particles increase in exactly the same way as Hubble had found for the intergalactic separations. In such a setting a "most non-chance configuration" is never going occur in the manner Eddington claimed.

Eddington's bald statement clearly relies on the tacit assumption, made unawares, that the particles are confined in one way or another. Indeed, for atoms in a box that collide randomly it is not unreasonable to assume the implications of both the recurrence theorem and the stronger ergodic hypothesis. However, the lack on Eddington's part of a qualification such as confinement to a box is surprising since he was well aware of Hubble's announcement of the law of expansion of the universe, made only two years earlier. In fact, he made the whimsical suggestion that his address was to a "mock Mathematical Association meeting", by which he meant one created by a fluctuation, and he even added a footnote in which he says he is "hopeful that the doctrine of the 'expanding universe' will intervene to prevent" the mock meeting being repeated many times over. This is simultaneously both more and less explicit than the 'loophole' which Boltzmann mentioned but did not follow up, which is that the universe might be infinite in extent and therefore not subject to the recurrence theorem: it is more explicit in referring to the expansion of the universe (of which Boltzmann had no inkling) but less so in not questioning the necessary condition for recurrence.

However, Eddington did contrast two possibilities, one of which seems to mark the first serious questioning of the Schuetz-Boltzmann idea and, with it, formulation of the 'Boltzmann brain' problem. Eddington said that a first *crude* assertion "would be that (unless we admit something which is not chance in the architecture of the universe) it is practically certain that at any assigned date the universe will be almost in the state of maximum disorganisation". He contrasted that with an amended assertion that "it is practically certain that a universe containing mathematical physicists will at any date be in the state of maximum disorganisation which is not inconsistent with the existence of such creatures".

Despite the fact that the probability for the amended possibility is clearly vastly less than for the crude possibility, Eddington considered it would still be so astronomically small that, in addressing the problem of the conflict between time-reversal symmetry and the second law of thermodynamics, he preferred "to sweep it up into a heap at the beginning of time", by which he meant what is now called the past hypothesis. Nevertheless, it marks a significant undermining of the fluctuation hypothesis as a serious explanation of observed time asymmetry applied to the whole universe. The hypothesis does not need to explain why the whole universe is the way it actually is; it only needs to account for the way we are deceived into thinking it is.

Note also that the expansion of the universe, which Eddington referred to in his footnote, is surely "something which is not chance in the architecture of the universe". The comparative recency of Hubble's discovery may explain why its relevance was not explored further. That cannot explain some much more remarkable later omissions.

THE FIRST WAS the ultimate refinement from Eddington's mock meeting of mathematical physicists to a solitary brain. It came only two years later in a little-cited paper by the Russian theoretical physicist Matvei Bronstein, a pioneer of the attempts to reconcile quantum mechanics with Einstein's theory of gravity, and his great compatriot Lev Landau. They noted that "a fluctuation would give rise to an observer without a surrounding world much more frequently than it would to the entire world known to us, so that no explanation is being offered for why we should be observing such a fluctuation". This is a splendidly succinct dismissal of the fluctuation proposal. It should have put an end to the Schuetz-Boltzmann hypothesis, but it didn't, as we will soon see. What is rather surprising is that in their paper the Russians, unlike Eddington, made no mention at all of the expansion of the universe, which by 1933 had been widely accepted as a fact. Should that not have called for a radical rethink of the application of statistical ideas to the universe?

We can see a remarkable example of the fact that it did not in Richard Feynman's discussion of the entropic arrow in the first volume of his *Lectures on Physics*, published all of thirty years later, in 1963; it does not include any mention of the expansion of the universe. After a beautifully clear account of Boltzmann's count-of-microstates definition of the entropy of gas confined in a box, Feynman considers the origin of the entropic arrow and says that

> one possible explanation of the high degree of order in the present-day world is that it is just a question of luck. Perhaps our universe happened to have had a fluctuation of some kind in the past, in which things got somewhat separated, and now they are running back together again. This kind of theory is not unsymmetrical, because we can ask what the separated gas looks like either a little in the future or a little in the past. In either case, we see a grey smear at the interface, because the molecules are mixing again. No matter which way we run time, the gas mixes. So this theory would say the irreversibility is just one of the accidents of life.

However, he argues that this is not a satisfactory explanation, for suppose "we do not look at the whole box at once, but only at a piece of the box. Then, at a certain moment, suppose we discover a certain amount of order. What should we deduce about the condition in places where we have not yet looked?" Note the argument continues in the setting of the box. It leads Feynman to say:

> If we really believe that the order arose from complete disorder by a fluctuation, we must surely take the most likely fluctuation which could produce it, and the most likely condition is not that the rest of it has also become disentangled! Therefore, from the hypothesis that the world is a fluctuation, all of the predictions are that if we look at a part of the world we have never seen before, we will find it mixed up, and not like the piece we just looked at. If our order were due to a fluctuation, we would not expect order anywhere but where we have just noticed it. . . .
>
> [But] the astronomers, for example, have only looked at some of the stars. Every day they turn their telescopes to other stars, and the new stars are doing the same thing as the other stars. We therefore conclude that the universe is not a fluctuation, and that the order is a memory of conditions when things started.

This observation, which Carl von Weizsäcker (a key figure along with Heisenberg in the German atomic bomb project) had already made in 1939, leads Feynman to say that the order is due not to a fluctuation but to a much higher ordering at "the beginning of time". He concludes: "This is not to say that we understand the logic of it. For some reason, the universe at one time had a very low entropy for its energy content, and since then the entropy has increased. So that is the way toward the future. That is the origin of all irreversibility". His final comment is that the matter "cannot be completely understood until the mystery of the beginnings of the history of the universe are reduced still further from speculation to scientific understanding".

In 1964 in lectures given at Cornell University and published in 1965 as *The Character of Physical Law*, Feynman made very similar comments, saying at the end: "Therefore I think *it is necessary to add to the physical laws* the hypothesis that in the past the universe was more ordered, in the technical sense, than it is today—I think that this is the additional statement that is needed to make sense, and to make an understanding of the irreversibility" (my italics). In a foreword to the 2017 republication of the book, the Nobel laureate Frank Wilczek says of the just-quoted passage that he finds Feynman's intuitive arguments convincing and that his "discussion of the origin of irreversibility frames the central question brilliantly, and takes a big step toward an answer". However, just as in the *Lectures on Physics*, the later book does not in any way mention the possibility that expansion of the universe could be centrally relevant. For all that, if the vanishing size of the universe at the Janus point does have the far-reaching consequences I mentioned as a possibility in Chapter 1, Feynman may have got near to the truth for an incorrect reason: near the truth because the conditions at the birth of time were very special indeed, but for the wrong reason because nothing needs to be added to the physical laws.

THE SAME CURIOUS lack of the universe, or rather its expansion, as a factor in the origin of the second law can be found in Penrose's books, beginning with *The Emperor's New Mind*. Of course, the expanding universe does often feature in his works, but not in direct connection with the statistical underpinning of the second law. Without citing Feynman, with whose discussion he may have been unfamiliar, Penrose arrives at much the same conclusion and actually goes further than him in a conjecture about a special condition at the big bang.

I will leave discussion of Penrose's proposal to a more appropriate point in the book. Here I merely want to mention the persistence of the 'box setup' in the argument that prompts it. When discussing Thomson and Boltzmann's discussion of atoms and molecules in a box, I commented how Penrose's beautiful diagrams clarify their rather terse arguments and readily persuade one of the overwhelming probability that any deep dip in entropy will have an increase to maximal-entropy equilibrium on both sides. I also commented that Boltzmann, unlike Thomson, stated that what happens in a box has an identical parallel in the universe: should the entropy of the universe at any instant have a low value, it will increase not only in the temporal direction of the future, as the second law requires, but also in the direction of the past, in flagrant violation of the second law. For much the same reason as Feynman, Penrose argues that this conclusion is manifestly refuted by our observation of the actual universe and, like Feynman, infers from this that the universe must, for some reason unrelated to statistical mechanics, have had an exceptionally low entropy in the past.

The point I wish to make in connection with the essentially identical arguments of Feynman and Penrose is that they hold only if the phase space of the universe has a bounded measure. Neither discusses this critical condition, without which the natural state of the universe as a whole cannot be one of thermal equilibrium. However, a footnote (on page 315 of *The Emperor's New Mind*) makes it clear that Penrose does assume the recurrence theorem will hold, because he says "if we wait for long enough" tiny low-entropy regions of phase space will be visited. However, he dismisses the relevance of this possibility on account of the "ridiculously long timescales" that are involved. This implicit recognition that the phase space of the universe has a bounded measure is also reflected in a calculation (on page 344 of the same book) that Penrose makes under the assumption that the universe will not only begin with a big bang but end in collapse to a big crunch. As I said earlier, I will return to Penrose's ideas, which have had a considerable influence in theoretical cosmology, later in the book. I have included them in this chapter to illustrate the now more than century-long influence of the Schuetz-Boltzmann fluctuation idea.

Giving something a name is important; in 2004 Andreas Albrecht and Lorenzo Sorbo, in a paper discussing cosmological inflation—the idea that the universe underwent a period of extremely rapid expansion immediately after the big bang—claimed that "inflation is the first idea with a chance to resolve the so-called 'Boltzmann-brain paradox'". The name has stuck. Would Boltzmann be pleased to know that his brain plays a hobgoblin's role in modern cosmology? I doubt it. Predictions that can be properly verified are the touchstone of good science. The critical question in any theory based on a multiverse is the relative number of universes like the one we observe ours to be. If the multiverse contains far fewer real universes like ours than disembodied brains, each with the belief it lives in a universe like ours, genuine predictions become impossible.

A remarkable feature of all post-Boltzmann discussions of the origin of time's arrows that I have consulted is that, with a single exception, none mention the critical role of Poincaré's recurrence theorem in making the existence of Boltzmann brains at all plausible. The exception is cited in the bibliography and is by the historian and philosopher of science John Norton. He gives several good reasons for their nonexistence besides failure of recurrence in a phase space of infinite measure.

It's now time for me to argue how real brains, yours and mine included, might be made by genuine creation. What might be the overall structure of a theory in which that can be done in a convincing manner? Staying faithful to Boltzmann's conviction, we will want not only the fundamental equations of the theory to be time-reversal symmetric but also all the solutions that it allows, or at least the great bulk of them, to share the same property. This is certainly the case in the Schuetz-Boltzmann fluctuation proposal. But, as we have seen, that proposal fails as an adequate explanation of our observations. An adequate explanation is one that needs no fine-tuning or artificial contrivance to get virtually all solutions of the theory to have bidirectional arrows of time not just for brief periods in small regions of the universe but throughout it and for all times in each of the oppositely pointing arrows.

Strangely, the first clear statement—that the problem of time's arrows requires a solution along these lines—seems to have been made, together with a proposal for one such theory, for the first time by Jennifer Chen and Sean Carroll in a paper in 2004 that I cite in the bibliography and discuss along with a further (unpublished) idea of Carroll's in the notes to Chapter 6. Both are clear anticipations of the basic structure of the kind of theory I propose in the remainder of this book.

CHAPTER 6

THE JANUS POINT

I do not know what I may appear to the world, but to myself I seem to have been only like a boy playing on the seashore, and diverting myself in now and then finding a smoother pebble or a prettier shell than ordinary, whilst the great ocean of truth lay all undiscovered before me.

—ISAAC NEWTON

ONE THING THAT THE READER SHOULD HAVE SEEN BY NOW IS MY SURPRISE at the absence, in all the discussions of time's arrows since the tussle between Boltzmann and Zermelo, of consideration given to the role played by confinement. Carnot's idealised heat engine and its conceptualised steam box went on to provide the perfect framework for understanding the nature of heat and a great deal more. But when it came to the ultimate explanation of time's arrows, I can't help feeling that too many theorists were, often unconsciously, looking in the wrong direction—into the box and not out into the universe. Surely there's an ocean of truth out there.

This book begins with Muriel Rukeyser's "The Universe is made of stories, not of atoms". Now, any one of Boltzmann's microhistories is a story of how molecules in a box dance around. It is true that all solutions, like the underlying laws, respect time-reversal symmetry, but does anything of interest happen? No. In a Poincaré-recurrent system, microhistories go on forever and at no stage do anything remotely interesting. Even the rare larger fluctuations lack structure. In a review of Samuel Beckett's two-act *Waiting for Godot*, the Irish theatre critic Vivian Mercier said it is "a play in which nothing happens, twice". The microhistories of a system trapped in thermal equilibrium are stories in which nothing happens, not just twice but infinitely many times. Interesting things happen only if such a system is acted on from outside. But nothing acts on the universe. How does it come about that not just interesting but downright fascinating things do happen in it?

The answer that I now begin to propose employs the simplest of models.* It uses what is called the *N-body problem*, which has a venerable history that began with Isaac Newton and remains to this day an active field of investigation. It concerns the behaviour of a finite number *N* of mass points subject to the forces of universal gravitation. This model will take us surprisingly far, indeed to the far-reaching consequences mentioned in Chapter 1 of a Janus point at which the size of the universe vanishes; Einstein's general theory of relativity and quantum mechanics will come later. My justification for using the *N*-body problem is that entropy and its increase were both discovered in an equally simple model: an ideal gas in a box. Several times now I have suggested that Boltzmann's failure, and that of many others, derives from unconscious retention of a 'box scenario'. Richard Feynman's is a remarkable example. My strategy is therefore this: to seek a model as simple as an ideal gas but without a confining box. The *N*-body problem is well suited for this purpose.

There is one deficiency of the model that I need to mention and explain why it is not fatal. In cosmology it is widely believed that the universe must satisfy what is called the *Copernican principle*.

---

* The only two models I am aware of that anticipate the model I now propose were mentioned at the end of Chapter 5 and are discussed in the notes.

This is a far-reaching and logical extension of the dramatic downgrading of the earth's status that began when the earth ceased to be regarded as the centre of the universe and became just one of the various planets that circle the sun. Astronomers now know that our galaxy, which hosts the sun, is just one among billions of similar galaxies distributed out as far as observations can be made. This fact suggested the Copernican principle: no location in the universe is distinguished in any outstanding way compared with any other. As yet, no theory can prove from first principles that this must be so, but a theory in clear disagreement with the principle can hardly be taken seriously. This seems to kill the use of an *N*-particle 'island universe', floating as it were in infinite space with its uniquely distinguished point at the centre of mass of the particles. Despite this violation of the Copernican principle, the model is still useful for two reasons: first, if the number of particles in the model is great enough, there are circumstances—and they are the critical ones—in which the particle distribution in the centre of the system is rather uniform and, locally at least, does look like a Copernican universe; second, the size of the particle model can increase and thus share the most striking feature of the real universe, its expansion.

The *N*-body problem brings with it a further significant advantage compared with models in conventional statistical mechanics. Any comprehensive discussion of entropy must take into account gravity, which has properties characteristically different from those of the other forces of nature. What I will do in the next few chapters is show how far we can get with Newton's theory, which has some remarkable properties that seem to be unknown to many of the theoreticians interested in time's arrows.

In fact, Thomson, whose 1852 paper first drew attention to irreversible processes in nature, did use Newton's theory to explain the origin and history of the sun and the earth. However, to do that he had to invoke a special initial condition: material particles far from each other at rest in a nebulous cloud. For its time, this was a credible theory and, with the all-important nuclear energy included, has the same thermodynamically dictated broad structure as modern theory. However, Thomson provided no explanation for the initial condition. He put it in by hand, just like one of the two proposals to which Boltzmann was finally driven to explain entropy increase—that the universe began in a special state for which no dynamical explanation can be provided.

In this chapter I will show how a Newtonian particle model has the potential to explain time's arrows. The model has what is effectively a special initial state, but it is not imposed ad hoc as the only way to make sense of irreversibility. The law of the universe, the overarching law of which I spoke in Chapter 1 (its structure will take shape bit by bit in the next few chapters), dictates its presence; it must be there. Gravity is important in determining its properties, but even more so is the freedom that the universe has to expand. That's what allows things much more interesting than anything which can happen in a box.

Chapter 7 will introduce refinements to obtain a more plausible model of the universe with architectonic features that match key structures in general relativity. These will then enable me to show the striking features such a model predicts; they are broadly concordant with what we observe in the real universe. That's the outline of what now follows. I will make no attempt to take quantum mechanics into account in any fundamental way until later in the book, and then only tentatively in connection with the implications mentioned in Chapter 1 of a vanishing size of the universe at the Janus point. There is still far too much mystery surrounding the interplay of gravity and the quantum, above all in the context of the universe. But ever in my mind is the conviction that true unity is to be found nowhere except in the universe comprehended as a whole. I see more than a little evidence for that—and will point it out—in what can be found in a purely classical theory of the universe.

Indeed, we find it quite soon. Newton became famous for relating the fall of an apple to the motion of the moon around the earth and for explaining why it is that planets orbit the sun in ellipses. However, these problems involve only two bodies and are not adequate for my purposes. I need at least three bodies. In fact, Newton did study one such problem, that of the motion of the moon in the combined gravitational field of the sun and the earth. He found this *three-body problem* fiendishly difficult and complained it gave him headaches. However, in 1772, the great Italian-French mathematician Joseph-Louis Lagrange won a prestigious prize with his study of the problem. Among many remarkable things, he obtained what is called a qualitative result—a first in the history of dynamics. What that means is best explained by saying what it is.

HERE AT LEAST some preparatory mathematics is hard to avoid; I will now do what I can to give you an intuitive feel for the first of three key quantities—there are only three—I will need in this book. Suppose children are playing on a school sports field during the class break. How might we characterise the size and distribution of the class, not simply by the number of pupils but also by how far they are apart and by their mass? If they all had the same mass and were uniformly distributed over the field, we might just say the size is measured by the distance across the field and by the number of the children. However, we might want a more refined characterisation that reflects the distribution of masses and whether the children are crowded together in one part of the field and spread out in another.

There are many ways in which this can be done, but one of them takes into account all the children and their masses in a simple and sensible 'democratic' way. Ever since Descartes introduced the idea, mathematicians have distinguished things—be they children, apples, or what you will—by letters of the alphabet. So let's call the children a, b, c, . . . and say a has the mass m_a and similarly for the whole class. Further, let r_{ab} be the distance between a and b. Of course, while this distance is a bit ill-defined for children, it is not so for mass points, and it is a good approximation when the children are far apart compared with their size.

We are looking for a number that defines an average distance and takes into account the masses, or rather their ratios. You will recall, I hope, the importance I attached to ratios of distances back in Chapter 1. This is because the aim is to model the whole universe, and there cannot be a ruler outside the universe to measure its size. This means we cannot talk sensibly about the size of a universe of just two particles; the ruler to measure its size would be the distance between the two particles, but that is the only thing that is to be measured. A ruler cannot measure itself. The size would always be 1 and nothing meaningful could be said about how the two-particle universe changes. But as soon as we have three particles, things are quite different. Three particles form a triangle with three sides, the ratios of which are well defined. However, to describe Lagrange's result, I will for the moment allow 'absolute distances', though in a form that can later be eliminated. What I can do conveniently already is employ only ratios of masses.

In whatever form the distances enter the expression we seek, the rules of geometry suggest the square root of a sum of squares will be involved. Pythagoras's theorem, the jewel in geometry's crown, says that the square of the length of the hypotenuse of a right-angled triangle is equal to the sum of the squares of the lengths of the other two sides. This suggests we consider the quantity

$$\ell_{\mathrm{rms}} = \frac{\sqrt{m_1 m_2\, r_{12}^2 + m_1 m_3\, r_{13}^2 + m_2 m_3\, r_{23}^2}}{m_1 + m_2 + m_3},$$

which is called the *root-mean-square length* (note the subscript $_{\mathrm{rms}}$). The presence of the sum of the masses, $M = m_1 + m_2 + m_3$, in the denominator means that ℓ_{rms} takes into account the masses democratically: the larger the mass, the greater its contribution to ℓ_{rms}. It's like a wealth tax in which each person pays tax according to their means. The same goes for the inter-particle distances (though they do not yet appear as ratios). Two large children on the sports field count more than two equally separated small ones, but they can rectify that by moving sufficiently far apart.

If there are N children or mass points, the number of pairings is $N(N-1)/2$. To see that is going to be the case, consider four children *a*, *b*, *c*, *d*. The pairs are *ab*, *ac*, *ad*, *bc*, *bd*, *cd*; these are six in number, and $6 = (4 \times 3)/2$, and so it goes on. Mathematicians express important quantities in ways both concise and beautiful. For the root-mean-square length of any arbitrary number of particles they write

$$\ell_{\rm rms} = \frac{1}{M}\sqrt{\sum_{a<b} m_a m_b r_{ab}^2},$$

where M is again the sum of the masses and the identifying letters a and b are understood to be numbers like the ones in the three-particle expression. The summation symbol Σ, the uppercase form of the Greek letter sigma, means add up all the quantities that follow it as restricted by the condition $a < b$, which means that a is to be less than b, so that distances are only taken into account once (the distance between two particles is a shared property, whereas each particle has its own mass). In other words: take the products of all pairs of masses and the corresponding squared inter-particle distances, add all the resulting expressions, take the square root of the sum, and finally divide by the total mass to obtain $\ell_{\rm rms}$.

I hope readers who struggle with mathematics have coped with this. If you do find it difficult, just think of the children, some small and some not so small, running around the sports field. You won't go far wrong in taking $\ell_{\rm rms}$ to be the diameter of the region they cover.

LAGRANGE'S RESULT, WHICH holds for any number of particles with arbitrary masses (the proof, given in the notes, requires only a few lines), concerns the behaviour of $\ell_{\rm rms}$. The result follows from Newton's laws of motion and the properties of the Newtonian gravitational potential energy, denoted $V_{\rm New}$. This is the second of the three key quantities, so I'll define it here and say just a little bit about it. You may recall that it was William Thomson who as a young man put the concept of energy and its two complementary aspects—kinetic and potential—at the heart of physics, replacing the archaic *vis viva* and attraction through space still used by Joule. In many cases, the potential energy depends on the distance between objects that attract or repel each other as Joule had in mind. For two particles of masses m_1 and m_2 with distance r_{12}

between them, it is proportional to m_1m_2/r_{12}. For a large collection of particles, it is

$$V_{\text{New}} = -G\sum_{a<b}\frac{m_a m_b}{r_{ab}},$$

where G is the Newton constant, about which I will say something in a moment. As you see, V_{New} is obtained by adding up the contributions of all pairs but now divided by the distance between them. The Newtonian force of attraction between any two particles, which changes their motion and with it their kinetic energy, is obtained by differentiation, if you know what that means; if not, it simply means that the force, in contrast to the potential (proportional to the inverse of the separation), is the famous inverse square law and equal to Gm_am_b/r^2_{ab}. The minus sign in V_{New} ensures that the force is one of attraction, not repulsion. The negative potential energy of the pendulum bob I used to illustrate the ebb and flow of energy conservation is simply the potential energy of the earth (assumed to be at rest) and the moving bob; its negative value gets less in absolute magnitude as the bob rises to a position further from the earth. Strictly speaking, potential energy is defined only up to an additive constant, often set to zero when the bob is hanging down. For gravitating particles in space, the expression above matches the common convention that $V_{\text{New}} = 0$ when they are all infinitely far apart.

It will help here to say something about the Newton constant G to underline the importance of ratios. Quantities that are not ratios are said to have dimensions. For example, the root-mean-square length ℓ_{rms} has the dimension length, expressed through square brackets as $[\ell_{\text{rms}}] = \text{l}$ ('ell' not 'one'); for the mass M we have $[M] = \text{m}$, meaning it has the dimension mass. The third fundamental dimension is t, for time. Pure numbers in the form of ratios have dimension [0]; for example, $[m_1/(m_1 + m_2 + m_3)] = 0$. The two sides of any formula that equates two quantities must have the same dimensions, which moreover must be defined relative to the same units. There was the space probe which crashed on Mars because one set of engineers were using the metric system and another imperial units. The dimensions of the Newton constant are normally determined by the way in which it appears in Newton's second law of motion, in which it relates the gravitational force to a resulting acceleration, making its dimensions expressed as $[G] = l^3t^{-2}m^{-1}$. However, defined simply in terms of mass and distance, it has dimensions m^2l^{-1}. Finally, anticipating the elimination of length in the description of the universe, you may have noted that both ℓ_{rms} and V_{New} contain lengths, the latter in its denominator. When the two quantities are multiplied together, we obtain a quantity with no length dimension; the mass dimension can also be eliminated by division by a suitable power of the total mass. That will give us a pure number, the third of the three key quantities I am going to ask you, as von Neumann recommended, to get used to.

That's enough mathematics and physics for a tea or coffee break.

LAGRANGE'S QUALITATIVE RESULT, the property he found when studying the three-body problem as a model of the earth-moon-sun system, requires one relatively mild condition to hold. It is this: the total energy of the particles—that is, the sum of their kinetic and potential energies (which by the law of energy conservation remains constant)—must not be negative. It must be either exactly zero or positive. It might be argued that this mild condition has the effect of creating a special initial state by adding, as Feynman advocated, something to the physical laws (that is, as a past hypothesis). However, if the special initial state is enforced by the law of the universe and the very laws of nature are themselves emergent consequences of that law, the argument fails. Whatever may be the reason, grant that the energy is not negative. Then the square of the root-mean-square length ℓ_{rms} when plotted as a function of time, as in Fig. 2, always takes one of the two forms that are shown: a U-shaped curve that is concave upwards and either passes through a finite minimum value, as on the left, or actually becomes zero at the minimum, as on the right. This latter possibility, which is exceptionally interesting, was not considered by Lagrange and there are difficult mathematical issues in saying that the square of ℓ_{rms} does go through zero and 'come out the other side'. It certainly can reach zero, and that will be sufficient for the more speculative discussion later in the book. (The plot shows the square of the root-mean-square length because it is proportional to what is called the centre-of-mass moment of inertia, which is the dynamically significant quantity.)

All that we need at the moment is what Lagrange did prove. Provided ℓ_{rms} does not vanish, there is a unique point at which it passes through a minimum value, and in both directions of time away from this point the size of the system, measured by the root-mean-square length, grows to infinity. The point of minimum size is clearly special and is the reason why, as explained in Chapter 1, I call it the *Janus point*. If the energy of the system is negative, three possibilities can be realised. First, if there are only two particles, they will never escape from each other: they will orbit their common centre of mass in elliptical orbits in the same way that, as Kepler discovered, the planets orbit the sun. Second, if there are three or more particles, they can all remain close to each other, either for a very long time or, exceptionally, forever. Third, and typically, the system breaks up very soon, but without the clean behaviour with steady increase of ℓ_{rms} shown in Fig. 2; instead of there being a single Janus point there will be a 'Janus region' of fluctuating ℓ_{rms} values before steady increase commences in both time directions.

Important in this connection is that the arrow on the time axis in Fig. 2 used to indicate the direction of time is purely nominal if

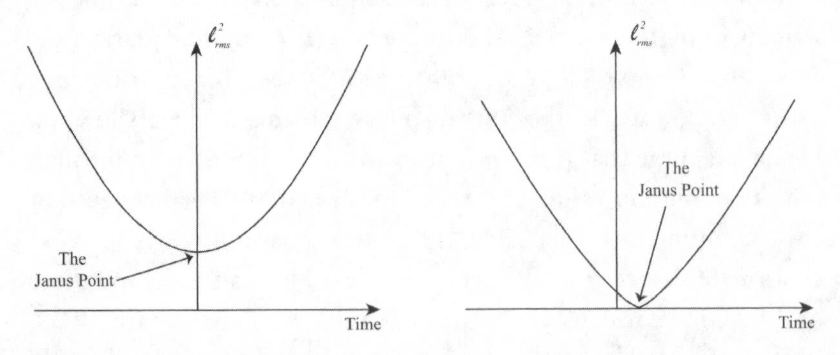

FIGURE 2. Two possible behaviours of the square of ℓ_{rms} when the energy is non-negative.

the three particles are taken to model not the earth-moon-sun system but the whole universe, for which it could just as well point in the opposite direction. As I emphasised in Chapter 1, films of divers emerging backwards out of a swimming pool only seem wrong because they confound the universal course of things as they unfold in the background of the universe. The background is simply not there if the three particles are the whole universe. Much of the book will explore the implications of this simple fact.

Lagrange's result is qualitative because it tells you just one single basic property of a whole class of solutions that satisfy a certain condition. Each individual solution will be different from every other except in this one common property. Besides its intrinsic interest and the fact that it was crucial for the development of all the ideas in this book, the universal property that Lagrange discovered was significant in the history of dynamics because it showed that if the energy is either zero or positive at least one particle of the system must escape from the others to infinity; otherwise ℓ_{rms} cannot grow forever. This means that such a system is unstable. It must break up. This realisation quite soon raised an important question: is the solar system, which has negative energy, stable? Is it possible that one day one of the planets will be ejected and fly off forever through interstellar space? In fact, it was for the solution of this very problem that in 1887 King Oscar II of Sweden and Norway offered a prize, the competition for which Poincaré entered and won with a study of the three-body problem. He proved as part of his monumental paper the recurrence theorem and, into the bargain, discovered chaotic behaviour—very sensitive dependence on initial conditions—in dynamical systems.

In Chapter 1, when I introduced the notion of a Janus point as a unique point of minimal size in the history of the universe, I suggested that arrows of experienced time point in opposite directions away from it along the timeline of the universe. The result obtained by Lagrange is the first example of such a point. It is very suggestive. To reflect that, I will call any dynamical system a *Janus-point system* if for it a quantity analogous to ℓ_{rms}—that is, a quantity that measures the size of the system—passes in all of its solutions through a unique minimum and increases without bound in both time directions. The unique point at which its size is minimal is its Janus point. It is important that the dynamical system is meant to represent the whole universe, not some collection of particles within it. They form a *subsystem* of the universe. The interest is in Janus-point systems as models of the universe. I will argue that the behaviour of subsystems, including their irreversible statistical properties, cannot be properly understood except in the context of the universe.

Observers within a Janus-point universe will necessarily be on one or the other side of the Janus point. It will appear to them to be a special point and, as I already said in Chapter 1, to be the beginning of time; I won't call it a big bang, reserving that for the case when the size is zero at the minimum. Moreover, as we will see, the dynamical evolution that such observers find around them will be characterised by a pronounced arrow of time from a birth of their universe at a finite time in the past to a never-ending future. Their universe will appear to be just as temporally lopsided as we find ours to be. What they won't know is what is granted only to a godlike observer with a bird's-eye view of the universe's complete timeline: arrows of time pointing in opposite directions in two halves of the complete Newtonian universe.* Despite temporal asymmetry in each half, time-reversal symmetry is respected overall, as Boltzmann clearly wanted.

The proposal I am making shares elements of both possibilities that Boltzmann thought might resolve the problem of time's arrow. He envisaged either violation of time-reversal symmetry through "unidirectional change of the entire universe from a definite initial state to a final state" or respect of the symmetry of the

---

*Timelines became very popular in the late eighteenth century as chronology—the science of ordering significant events in time—developed. The famous philosopher Henri Bergson strongly disliked the notion of a timeline because it 'spatialised time', whereas for him time and space were very different. This was why, like some physicists (including my friend Lee Smolin) and philosophers of science, he thought that time is fundamental, with a uniquely distinguished direction from past to future. He contrasted this with space, in which there are no distinguished directions. I'm indebted to Emily Thomas for the information about timelines and Bergson's objections to them.

complete universe but with spatially and temporally separated 'worlds' within it in which time's arrow points, at one instant, in opposite directions. In the proposal now being made, there is unidirectional change from what appears to be a special initial state, but in fact there is also unidirectional change on the other side of that point; time-reversal symmetry is respected, though not in the way Boltzmann imagined. Thus, I think the new proposal gets the best of both worlds.

Apart from the two anticipations (not in fact connected to Lagrange's discovery) that I mention at the end of Chapter 6, I am not aware of any recognition prior to a paper that I and two of my collaborators published in 2014 that behaviour of the kind shown in Fig. 2 might have the potential to resolve the mystery of time's arrows (the idea first occurred to me in the spring of 2012). The reason for this may be the specialisation that is now so characteristic of science. Lagrange's result, which is actually only a minor part of his 1772 paper, is well known to *N*-body specialists, but generally speaking they are not interested in the origin of time's arrows and the growth of entropy. For their part, the theoreticians who have grappled with the issue of time's arrows seem to have been unaware of the behaviour shown in Fig. 2.

I should warn the reader against a not altogether helpful analogy that has gained some traction in discussions of the Janus-point idea in the media, especially online: namely, universes that 'bounce'

FIGURE 3. Janus as depicted on a Roman coin. Wikipedia Commons: https://commons.wikimedia.org/wiki/File:Janus_coin.png

near the big bang from a contracting to an expanding phase. Such models have often been developed to avoid the universe having zero size at the big bang, which is held to be scientifically unacceptable because key physical quantities become infinite. In other cases, the 'bounce' models aim to explain currently observed properties of our expanding universe through properties that existed in the earlier contracting phase. The analogy is with the bounce of a ball: the direction of motion is reversed or changed abruptly while time advances, quite unaffected. However, a Janus-point universe is not like that. A much better illustration for the situation shown on the left of Fig. 2 (the alternative possibility on the right is discussed in Chapter 16) is with a ball dropped into a hole bored through the earth to the antipodes. As the ball approaches the centre of the earth there is no bounce. Instead, the ball's velocity actually stabilises for an instant because there it is not subject to any gravitational attraction to the centre. What does happen, imperceptibly for anyone on the ball, is a reversal in the direction of the vertical. This is analogous to reversal of the direction of time at the Janus point.

It should also be noted that another online misrepresentation, exact mirror reflection, is not present. The fine Janus image on a Roman coin shown in Fig. 3 was found online with the caption "Janus, god of beginnings" and is particularly apt as an illustration of the absence of mirror symmetry between two half-universes that are only qualitatively similar. You will see that the two faces are slightly different, not mirror images turned back to back. The Roman craftsman who minted the coin in Fig. 3 did a superb job but obviously could not make the faces identical.

I call the model that in Chapter 7 illustrates the potential of the Janus-point idea the *minimal model*.*

*Nigel Goldenfeld coined the term (see "Minimal model explanations" by R W Batterman and C C Rice in *Philos. Sci.* **81** 349 [2014]). I changed to that from 'toy model' (used by physicists for a greatly simplified model of an actual system) on the suggestion of Michael Berry, who says 'toy model' has for him "always carried a half-apologetic disparaging aroma of triviality". I think the implications of the minimal model are far from trivial.

CHAPTER 7

THE MINIMAL MODEL

IN THE NEWTONIAN DYNAMICS OF ISOLATED SYSTEMS—ONES NOT SUBJECT to external forces—three quantities are especially important. This is because their values are *conserved*, remaining unchanged as the system evolves. The most important of these quantities has already played a central role in our story and came to great prominence with the discovery of the first law of thermodynamics; it is energy. The momentum of a particle moving freely is conserved, as is the total momentum of a system of particles that interact with each other. I therefore only need to say something about the third quantity, angular momentum, which is a measure of the overall rotation of the matter in the system. The illustration often given of its conservation is a ballerina pirouetting on a toe. Every part of her body is rotating about the vertical line through the centre of her body. If she begins the pirouette with her arms outstretched but then draws them in, the pirouette becomes significantly more rapid. It must because what counts is the speed of rotation of each part of her body multiplied by its distance from the central vertical. When she reduces that by pulling in her arms, the speed of rotation must increase. The effect is beautiful.

I will have more to say about conserved quantities in Chapter 8, but for now, in describing possible solutions of the three-body problem as a minimal model, I will consider only those for which all three conserved quantities are exactly zero. In fact, vanishing momentum is not a restriction. By Galileo's relativity principle, as also generalised by Einstein, the total momentum of any isolated system has no effect on the processes that unfold within it; it merely reflects the motion of the system's centre of mass. In contrast, exact vanishing of energy and angular momentum is an extremely strong restriction. Examples corresponding to it can't be found in nature around us. Nevertheless, I believe we should consider, as models of the universe, only those for which the conserved quantities are exactly zero. This, in a step that will be justified in Chapter 8, will add to the mild condition that the total energy is non-negative, and take us much further toward what I conjecture is the most perfect imaginable form the law of the universe could take. The final step will be to impose the condition of vanishing size. Without this last condition and one other possibility, both of which lead to very special solutions, all possible solutions that we need to consider at this stage have the form illustrated in Fig. 4. Please have a look at it now; everything in the rest of the book builds on it.

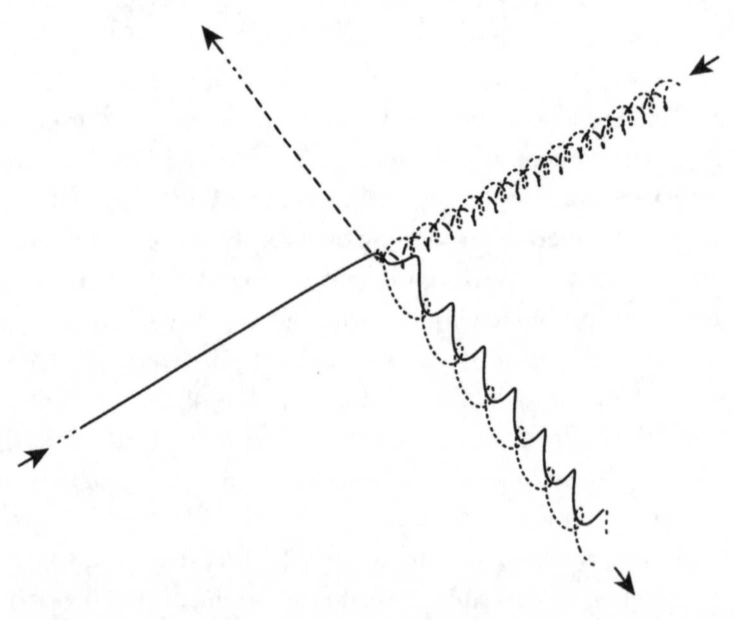

FIGURE 4. A typical history of the minimal model.

It is helpful that all solutions with vanishing angular momentum are planar—the three particles remain throughout in a fixed plane. This makes it possible to depict the paths they take as curves on the page. In the distant past, one particle, the *singleton* (solid curve), comes (from the bottom left in the figure) towards the other two particles (dashed and dotted curves). These two are orbiting their common centre of mass in Keplerian elliptical motion. Initially, gravitational forces play a significant role only in this *Kepler pair*, but as the singleton gets closer, the gravitational forces between all three particles become significant. The motion becomes chaotic and hard to predict. The size of the system, measured by the root-mean-square length ℓ_{rms}, passes through its minimal value at the Janus point. No matter what happens in that region, the outcome, apart from the very special solutions, is always that the system breaks up again into a singleton and a pair. There may, as in the illustrated example, be a swapping of partners, the previous singleton becoming the partner of one of the two members of the previous pair. If we invoke the time-reversal symmetry of Newton's laws to reverse the direction of time and motion indicated by the arrows, the qualitative nature of the evolution is the same. As I emphasised in Chapter 6, the ambiguity arises because the three particles are taken to represent a universe, not a subsystem within it. The conceptual difference is critical.

There's a way of thinking about the manifestation of time-reversal symmetry in this example that highlights the puzzle it presents. Suppose the singleton is not a mass point but a young man looking for a partner as he walks across a vast ballroom floor. In the distance he sees a couple waltzing towards him. When they get close enough, he takes his chance, grabs the young woman, and dances off with her. The jilted partner is left to walk away disconsolate in the opposite direction. But if we reverse the direction of the purely nominal arrows, he's the one who gets the girl. Bizarre interpretations like this never enter our minds in the real world because our lives unfold within the overwhelming background of what is happening in the world and, above all, in the universe at large. If two billiard balls collide on a table, two alternative 'directions of time' can always be formally associated with the collision. This is never done because we take the direction of time from everything else that is happening. But for our minimal model that possibility falls away. There is nothing by means of which we can define a direction of time except what we find within the system itself. It would seem we are left with nothing to define the direction of time. Increasing the number of particles does nothing to resolve the dilemma.

THE RESOLUTION OF the dilemma that I am going to propose has similarities to the pragmatic idea to which Boltzmann found himself driven when trying to explain why entropy increases with time. In his case, necessity was not so much the mother of invention but a gratifying insight. It is wrong to seek to explain why the direction of entropy increase coincides with the direction of time. The direction of entropy increase *is* the direction of time. A single direction had been mistaken for two. I'll comment on the similarities and differences between his and the Janus-point proposal after I've presented it. It is based on the way the size and, more importantly and much more interestingly, the shape of the triangle formed by the three particles change in the various stages of the evolution.

Consider first the size as measured by the root-mean-square length ℓ_{rms}. In the sequence indicated by the arrows in Fig. 4, it decreases steadily from the remote past to its minimum at the Janus point and then increases again all the way to infinity. If we seek to assign a direction of time, not by nominal stipulation but by what happens within the system itself, then a first possibility is to use size and say that there are two arrows, each pointing away from the Janus point in the two directions of increasing ℓ_{rms}.

However, this is not ultimately satisfactory. It presupposes something outside the system, an external rod, that measures its size. But the universe is everything. Anything we say about it must come from within; size is suspect. Observers within the universe can only measure angles and ratios of distances. They can determine the shape of the universe, not its size. Let's think about that. On the left of Fig. 5 the arrows have been removed—there is now

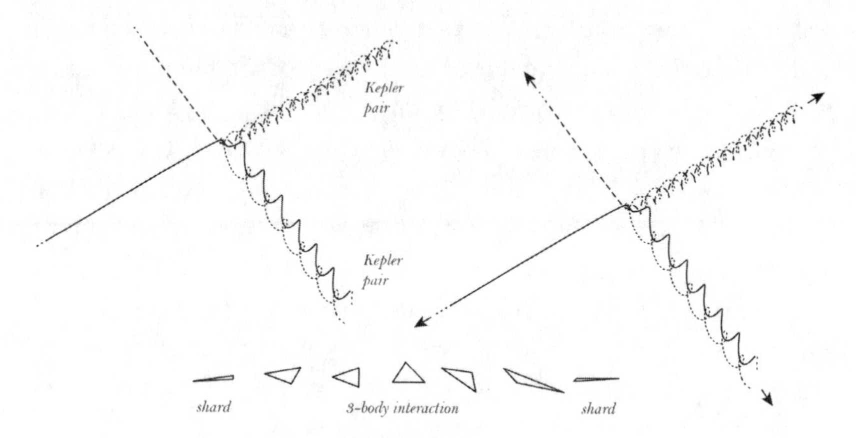

FIGURE 5. The one-past–two-futures interpretation.

nothing to suggest the passage of time—and the typical shapes of the triangles at the various stages of the evolution have been added. The shards are when the singleton (at the tip of the shard) is far from the Kepler pair's particles (at the other two vertices of the triangle), while the more nearly equilateral shapes are realised in the region of the three-body interaction. The two diagonals represent the two distinct stages of the evolution; in the conventional Newtonian interpretation, neither can be said to come before the other. The Gordian knot is cut on the right of Fig. 5, in which the arrows now point along both diagonals *away* from the Janus-point region of chaotic three-body interaction. There are now two effective histories within a single solution, one on each diagonal. Represented on a timeline, they become the back-to-back 'half-solutions' promised in Chapter 1. In each, a singleton separates itself from the other two and tends to motion that becomes progressively more rectilinear; at the same time, the other two settle down into a Kepler pair.

As regards the assignment of a direction of time to the universe, we will obviously want it to be objective. In fact, I have already prejudged the choice on each diagonal on the right of Fig. 5, but let's consider the criteria that should be applied. The most important is that it must always be the same if chosen by two judges not allowed to communicate with each other. If asked to give a temporal order to the complete line of triangle shapes in Fig. 5 for many different solutions, each judge would simply be forced to make a random choice. There would be agreement at best about 50 per cent of the time. In contrast, even small children could look at the line and see clearly that the shape evolution in the central region is very different from what develops in the two 'wings', where the triangles become shards. Given this characteristic change in shape, and allowed to assign arrows not for the complete line but for parts of it, both children and Nobel laureates would point the arrows in opposite directions in the two halves: either towards or away from the central region. They would assign bidirectional arrows of time.

This still leaves open an important question: can we identify past and future in each half? This is ultimately a pragmatic question. Already in the three-body problem we see that, in each diagonal, the solutions become more regular and structured going away from the central region. We will see that this effect becomes much more pronounced when there are many particles and not just three. Moreover, the motion is chaotic near the Janus point. This and further arguments to be given later will clinch the case for having the arrows point away from the central Janus-point region. We will have a situation with one past and two futures.

In introducing bidirectional arrows, we will of course be matching Boltzmann's use of entropy growth to define the direction of time in systems subject to fluctuations on account of Poincaré recurrence. But Boltzmann's arrows point from order to disorder; ours, in the model in which the size remains finite at the Janus point, point from disorder to the creation of structure (structure creation happens in both cases; the only difference is at the Janus point). The origin of the difference from the Boltzmann scenario is that Janus-point systems are not subject to recurrence—they are not in a 'box'. Moreover, the arrows exist not because of statistical fluctuations but because of dynamical necessity. The law of the universe dictates their presence. Also there are infinitely many entropy dips in a confined Poincaré-recurrent system but just one in a Janus-point system. In fact, the dynamical origin of the Janus point leads to effects that can be characterised much more sharply than the rather vague order/disorder contrast on which Boltzmann relied.

CLEAR SIGNS OF this can already be found in the present model. I noted earlier that the size of the universe cannot be defined by an external measure. The same applies to time. It cannot come from a clock outside the universe. If we are to find time within it, we need to distinguish *chronology*, the ordering of events, and *duration*, the time between them. We have seen that our model universe's shapes, the realisation of which are the *events* in its history, can be arranged in a linear order with 'before' and 'after' defined on each side of the Janus point. Chronology does not present problems.

Considering now duration, we find a very interesting effect. As we move away from the Janus point in either direction, the first significant thing is that the system breaks up into a singleton and a pair of particles. At this initial stage, the motion of the two particles around each other is not regular enough to be called Keplerian. Indeed, it never becomes perfectly so because the motion within the pair is always perturbed to some extent, however small, by the singleton no matter how far away it is. However, once the pair and the singleton are far enough away from the Janus point, the fact that the three particles always lie in one plane means that the system must pass repeatedly through instants of collinear configuration, when a straight line can be drawn through all three points. Let these instants be called the ticks of a clock and the separation between the particles of the pair at the ticks the length of a rod. This last can be used to measure, at the time of the ticks, the distance d from the pair's centre of mass to the singleton and determine how d increases as the 'time' measured by the ticks increases. The answer is easily guessed: d will always increase between each tick, initially rather irregularly, but with ever better uniformity as the pair's motion becomes ever more Keplerian.

In accordance with Newton's first law, a body subject to no forces will move forever with uniform speed in a straight line. Relative to the rod-cum-clock that the pair becomes, one can certainly say that as the number of ticks from the Janus point increases, the singleton moves with ever more perfect uniformity. We see the emergence of the uniformity of Newton's first law together with a composite 'instrument' which verifies that part of the law. The other part of the law, the rectilinearity, also emerges. As the particle pair becomes ever more Keplerian, the angle between the line along which the major axis of its elliptical motion points and the line to the receding singleton settles down to a fixed angle. The major axis becomes the 'needle' of a compass. With respect to the direction it defines, the singleton moves with ever better rectilinearity. The Kepler pair becomes an ever more accurate rod, clock, and compass all in one.

The overall picture is this. All of the minimal model's solutions respect time-reversal symmetry in having two qualitatively similar halves. Objective criteria permit assignment of arrows of time to each half that, judged by increasing regularity of the motion, point in opposite directions away from the Janus point. Whereas the complete solutions are symmetric, each half is, considered alone, pervasively asymmetric. This can be seen especially clearly in the way the law of inertia together with the rod, clock, and compass to confirm it emerge. This picture underlines the great difference between confined and unconfined systems. The motions of three particles subject to the same laws as the minimal model but trapped in a box with walls off which they bounce elastically could never settle into the motion described earlier. Any incipient formation of a Kepler pair would soon be disrupted. To come into existence, interesting regular motion needs space. The girl's footprints in the sand can trace out intricate paths like those in Fig. 5.

An observation worth making here concerns the conditions under which there is creation of 'precision instruments', by which I mean in the present context the formation of a Kepler pair as a rod, clock, and compass all in one. Apart from the very special solutions not yet discussed but which are just as striking, this acme of perfection is guaranteed to appear in what I will call a *relational universe* and show why in one both energy and angular momentum are zero. This statement is absolute. I conjecture that precision, realised 'in the fullness of time', is a characteristic feature of the law of the universe. If the conditions of vanishing energy and angular momentum are dropped, Kepler pairs can still be formed, but there is no guarantee that they will. In particular, if the energy is positive, the particles can sometimes simply fly off to infinity without forming a rod, clock, and compass all in one. In such a case they won't do anything interesting.

WHAT ARE WE to think about temporal language and our place in the universe if it does have a Janus point? If that is the case, we, as observers, must exist on one or the other side of that point and experience an arrow of time. We will be aware of a past, present, and future and believe that causes come before effects. All this is dictated by and emergent from a timeless law of the universe. Our arrows of time have a dynamical origin, not a statistical one. They may appear to rely on a past hypothesis, but they won't if the law of the universe has the form I conjecture. Already we only need to require the energy to be non-negative to ensure the existence of arrows in every single solution. With every further plausible condition we identify in the putative law of the universe, the arrows will get sharper and sharper.

St Augustine is famous for saying, "God did not create the universe in time but with time". That's what we find in the minimal model. It has an infinite timeline; there is no creation ex nihilo. But the universe is created by law with time. As regards the ultimate question of what stands behind existence and the law that governs it, Boltzmann would not hazard a guess, nor will I. However, there is a way I like to think about these things.

In words as famous as St Augustine's, the seventeenth-century philosopher Baruch Spinoza said that the most fundamental problems need to be seen *sub specie aeternitatis*—from the perspective of eternity. If we are to adopt such a perspective, it will surely help to have a notion of what eternity actually is. Is it made of something? How is it ordered? And what is our place within it?

My answer to the first question is that eternity is made of instants of time. And what are they like? At least in their calculations, many mathematicians and physicists identify instants of time with featureless points, all identical and, in the way that Bergson so disliked, strung out on a timeline. But are such points real things, as Newton claimed for both the instants of time and the points of space? Leibniz disagreed strongly. He said that a real thing must be concrete and have attributes. The word 'concrete' derives from the Latin *concrescere*, 'to grow together'. A thing must have parts that are held together. Leibniz liked to speak of a plurality within a unity. It's easy to grasp what he meant. Stand near a seashore where waves break, and open and close your eyes repeatedly. Consciousness keeps on presenting you with different patterns of water seen momentarily all at once and held together within your field of view. That's something concrete, freighted with variety. In models of the cosmos, instants of time should be like that.

In the three-body minimal model, each instant is a triangle shape with masses of definite ratios at its vertices. Geometry and masses are the quintessence of dynamics. Any triangle shape 'decorated' with mass ratios at its vertices is an instance of an instant. You can, as I do in my book *The End of Time* (but, except in Box 3, without removing size and leaving only shape), call it a Now. A child such as my five-year-old grandson Jonah, given all the triangle shapes of Fig. 5 in a jumbled heap, could sort them out and lay them in the correct order on a line. In a well-known passage in his *Slaughterhouse Five*, Kurt Vonnegut explains how his time-travelling Tralfamadorians comprehend time: like the Rockies seen all at once from a distance. Each peak of the Rockies is an instant of time. The Tralfamadorians see time *sub specie aeternitatis*. Philosophers of science characterise such an intuition, which I share, as eternalism. Because we are in the universe and have a powerful sense of the passage of time (which I will discuss at the end of the book), I will in the remainder of the book often use conventional temporal language. But, more or less following St Augustine, this should be taken to imply not creation of the universe in time but rather its eternal existence with time. If the ideas presented in this book are correct, the big bang is merely a special point on an emergent infinite timeline—emergent because a child could construct it from the shapes that the law generates. In fact, if it were then broken up and all the shapes put back in the jumbled heap from which it could have been constructed, it could, with a happier fate than Humpty Dumpty, be put together again.

Much as I enjoyed Vonnegut's book, especially the inspired human decency of the bombing of Dresden (which he had experienced and survived as a prisoner of war) run backwards in time, the Tralfamadorians' image of time is still inadequate in two respects.* First, mountains do not properly capture the way the shapes change in a pronounced unidirectional way with increasing distance from the Janus point. Second, the Rockies stand on the earth's crust; it gives them not only support but also a separation between their individual peaks. Some philosophers of science, who essentially follow Newton and accept his notions of absolute space and time, subscribe to substantivalism, in accordance with which space can be likened to an infinite translucent block of ice in which each point defines a place and lines can be drawn. Time is reduced to something like a straight wire—Bergson's spatialisation of time—along which distance between two points exists as duration before events occur at them. Space is a container of locations, time a container of events. The problem with space and time conceived this way is simple. They are invisible. We see things; we see neither time nor space. They are not our home; something else is.

The emergence of the rod-cum-clock in the minimal model suggests that the preexisting timeline is redundant. All we need to construct such a line, *a posteriori* and with back-to-back directions on it, is, in the simplest case, triangles endowed with mass ratios and governed by a law. As Chapter 8 will show, space—or, rather, absolute space as Newton conceived it—is also emergent.

---

*I bought Vonnegut's book at Borders on the Magnificent Mile in Chicago in November 2000 when promoting *The End of Time* at the city's Humanities Fair. I was able to show it to the audience of about five hundred and say that American reviewers of the book, which I had not yet read, had likened my ideas about time to Vonnegut's. It was only after my return home that I read the book and found that its antihero Billy Pilgrim, having time-travelled with the Tralformadorians, knew how he would die: by assassination on the occasion of a visit to Chicago to address a large public meeting on flying saucers and the true nature of time. I had better luck than Billy.

In fact, the task we face is twofold: to show how the conceptual world of classical physics emerges from a quantum law of the universe and, within that classical framework, to show how Newton's and Einstein's laws emerge. That second task is addressed in the chapter which follows. In a sense, in this book I am forced to put the cart before the horse because as yet no fully developed quantum theory of the universe exists. I therefore consider mainly the most fundamental aspects of classical physics, above all the emergence of the arrow of time, to which several chapters are devoted. Towards the end of the book I do make a tentative proposal for the form, entirely consistent with the basic ideas on which this book is based, that the quantum law of the universe might have. If it is correct, and of course it's a big 'if', it also amounts to what truly is a new theory of time. Here's a challenge for readers familiar with the main problems that have been thrown up by the decades-long search for quantum gravity: see if my idea occurs to you before you get to it.

CHAPTER 8

THE FOUNDATIONS OF DYNAMICS

The universe is given once only, with its relative motions alone determinable.

—ERNST MACH

Historical Background

When Ernst Mach, who gave Boltzmann that difficult time by questioning the existence of atoms, died in 1916, Einstein's obituary of him lauded his "incorruptible scepticism". This was not so much about atoms but rather about Newton's fundamental concepts. The motto at the start of this chapter is taken, with a slight rearrangement, from Mach's *The Science of Mechanics* (1883), which includes a sustained critique of Newton's notions of absolute space and time. Newton introduced both in order to define motion with mathematical precision. Let's consider space first.

Imagine a room. The distances from the three walls fix the position of any point within it. You know what it means to say the point moves in a straight line within the room. That's what Newton wanted from his absolute space. But it was a room without any visible walls. Newton only had something like the block of ice mentioned at the end of Chapter 7. This is just as problematic because the ice is invisible. You only see relative motions.

Of time Newton said that it flows uniformly "without relation to anything external and by another name is called duration". But in the famous experiment which established the law of free fall and laid the foundations of dynamics as a precise science, Galileo rolled balls down a smooth inclined plane and measured the time it took for the ball to traverse a certain distance by the literal flow of water from a tank: so much water, so much time. Facts like this led Mach to comment, "It is utterly beyond our power to *measure* the changes of things by time. Quite the contrary, time is an abstraction, at which we arrive by means of the changes of things".

Galileo's experiment is inconceivable without the ball, the board, and the water flowing from the tank. It involved motion relative to tangible things. Remarkably, Newton felt himself forced to banish them from his conceptual world. He was reacting to the final disintegration of the Aristotelian cosmos, which for two millennia had placed an immobile earth at the centre of a star-studded revolving sphere. Position was geocentric: defined relative to the solid earth. Copernicus had set the earth in motion and stopped the revolution of the heavens. The earth could no longer be taken to define position. But a key part of the Aristotelian cosmos survived: both Copernicus and Kepler took the sphere of the heavens, believed solid, to be by definition at rest and to serve as the ultimate frame relative to which rest and motion are defined. In their cosmos, everything had its place. With his *Principles of Philosophy*, published in 1645, Descartes demolished this reassuring conceptual home by replacing it with the mechanical philosophy. According to it, all phenomena in nature are to be explained by collisions of matter with matter. The rigid sphere carrying the star-studded heavens disappeared without trace as the closed world became an infinite universe filled throughout with matter in motion. Descartes's mechanical philosophy was very influential; greatly transformed and developed, it still is.

Newton was not averse to the mechanical philosophy but saw a profound problem in Descartes's claim that motion of any matter is always relative to other matter. That is certainly how we recognise motion, but Newton wanted the science of motion to match the rigour of Euclid's geometry. He wanted laws of motion from which he could deduce theorems. In a world in which everything moves relative to everything else, where can one find fixed markers to define motion? He could see no way in which one could identify points that are *equilocal*—occupying the same place at different instants. This was critical. Newton already had a sense of the great potential of the law of inertia, in accordance with which a body unaffected by forces could either move uniformly in a straight line or remain at rest forever at the same point. Cartesian relativity made a mockery of such a law if the same point could not be identified at different times.

Newton's solution was radical. Rejecting visible material markers, he introduced mathematical markers: identical, immovable, and invisible points of space all packed cheek by jowl in an infinite expanse. This was his absolute space; motion within it was absolute. Leibniz said absolute space was a fiction. Because it lacked all variety, nothing could identify or locate its points.

MACH HAD THE same intuition; he too, 150 years later, was adamant that absolute space could not be the foundation of dynamics. Science must be about things that can be measured. You cannot measure distance between invisible points, but you can between physical objects. Mach had an instinctive holistic conviction and felt the universe must be an interconnected whole. In 1883 he said, "Nature does not begin with elements, as we are obliged to begin with them. It is certainly fortunate for us, that we can, from time to time, turn aside our eyes from the overpowering unity of the All, and allow them to rest on individual details. But we should not omit, ultimately to complete and correct our views by a thorough consideration of the things which for the time being we left out of account".

Please note the final words. I have been arguing in the first part of the book, beginning with the perfectly understandable reason why Carnot did not seek an explanation for the existence of coal, that the origin of time's arrows is to be sought not in the here and now but in the furthest reaches of space and time.

For his part, Mach did try, with very inadequate mathematics, to formulate a dynamical law of the universe expressed entirely in terms of changes in the distances between all the bodies in the universe. This would dispense with Newton's invisible points of space. The key difficulty was to explain why Newton's laws worked so well; you certainly can find frames of reference in which, in accordance with Newton's first law, a body subject to no forces moves in a straight line at a uniform speed as measured by mechanical clocks. They are called *inertial frames*.

Galileo was the first to notice their existence; they formed a key part of his defence of the Copernican revolution, part of which implied that the earth must be spinning on its axis and, depending on the latitude, carrying bodies along at a speed people found inconceivable—about 1000 miles per hour on the equator and a substantial fraction of that at European latitudes. People argued that such speeds must have catastrophic consequences on the earth; as it rotated to the east, church steeples would fall backwards to the west and howling gales must blow in the same direction. To counter these arguments, Galileo pointed out that on a galley sailing smoothly relative to the land all known physical processes unfold in exactly the same way as on the earth whatever the speed of the galley. This phenomenon is called *Galilean relativity*. In fact, Galileo thought it applied to circular motion around the earth, but by Newton's time it was recognised that there exists a whole family of frames of reference, quite unrelated to the earth, in which phenomena unfold as on Galileo's galley. From the 1880s, they came to be called inertial frames of reference because in them bodies not subject to any forces satisfy Newton's first law of motion.

There are two rather remarkable things about these frames of reference: first, that they exist at all and, second, that there is a whole family of them which all move at uniform rectilinear speed relative to each other. Within each such frame, the conditions that Newton postulated for absolute space and time hold good. The existence of the whole family of frames, in each of which the same laws hold, is called the *relativity principle*. In the famous paper with which he created the special theory of relativity in 1905, Einstein assumed this principle applied not only to mechanical processes but to all physical phenomena. Coupling this with the simple assumption, subsequently well confirmed, that the speed of light does not depend on the speed with which the object that emits it is moving, Einstein deduced the remarkable behaviour of clocks and rods for which he is famous together with the equally famous equation $E = mc^2$.

Some twenty years before that, Mach had conjectured that, through some as yet unknown mechanism, the totality of masses in the universe creates the mysterious local inertial frames of reference through a powerful guiding influence. He observed that Newton's mathematical results were actually confirmed not relative to absolute space and time but relative to the frame of reference defined by the fixed stars and to time defined by the diurnal sweep of the stars across the meridian. Mach's rather vaguely formulated conjecture attracted much interest. In fact, it hugely influenced Einstein when he read Mach's book sometime around 1900. It became the primary motivation in his search for a new theory of gravitation that culminated, after great struggles, in the creation in 1915 of the general theory of relativity. This has proved to be an amazing success, most recently with the detection, exactly as predicted, of gravitational waves resulting from the merger of black holes and neutron stars and also the observation of the so-called event horizon surrounding a massive black hole in the galaxy M87. However, the extent to which the theory implements Mach's idea remains unresolved.

The reason is easily stated. Einstein did not construct his theory in a way that would show how, through a specific mechanism, the universe at large directly determines the existence of the inertial frames we find around us. Instead, he introduced a profound modification of them. In one of his most brilliant insights (the "happiest thought" of the footnote on page 44) he realised the inertial frames could not, as in Newtonian theory, have infinite extent but could extend only locally over regions within which the variation of the gravitational field can be ignored. The relativity principle still applies to *freely falling* frames, such as those realised in space probes falling at any speed in a gravitational field that is effectively constant over the probe dimensions. By analogy with Galileo's galley, Einstein required that the laws which govern all phenomena observed in freely falling frames should always take exactly the same form in any one of them. He called this the *equivalence principle*; it made gravity's effect locally indistinguishable from inertia's. Together with some further conditions, this sufficed to fix the fundamental equations of the theory. He required them to hold everywhere in its overarching concept, spacetime. It could have any structure in which at each point the equivalence principle holds. Each infinitesimal region satisfying this condition became a conceptual 'brick' and any spacetime a 'cathedral' built with many such bricks. Their form is to some extent flexible, allowing many cathedrals to be built from them. The one that describes our universe more than rivals Gaudi's Sagreda Familia in Barcelona, if nothing else because it made Gaudi and contains his masterpiece. But others are not so impressive; for example, the simplest of all, Minkowski space, is as bland as Newton's absolute space.

In fact, Einstein's achievement was not quite what Mach had had in mind. It was not to see how bricks can be used to build cathedrals but to understand how the bricks are made in the first place. Einstein left open the question of whether, and by what mechanism, the universe at large determines the inertial frames of reference observed locally.

In 1918, considering the structure of his new theory, he formulated a requirement on possible cosmological solutions of the theory under which one could directly see how the masses in the universe, in their totality, determine the local inertial frames of reference. It would then satisfy what he called *Mach's principle*. The name stuck, but it is important to distinguish Mach's original idea from what Einstein did with it. For very understandable historical reasons, Einstein did not embed Mach's idea directly and essentially into the foundations of his theory. In fact, as just illustrated with the bricks and cathedrals metaphor, he actually made the structures whose origin he wanted to explain—the local inertial frames of reference—the unexplained building blocks of his theory. The strategy he employed to create the theory succeeded for two reasons: first, the unexplained 'bricks' still represent a profoundly important fact about conditions everywhere in the universe, as no violation of the equivalence principle has ever been found; second, he brilliantly used the further well-founded conditions mentioned earlier.

However, as regards the Machian aspect, he got into a muddle. Without realising it, he was trying to fit his wonderful theory into a Procrustean bed. Near the end of his life he realised this and even said that it "will be immediately recognised" that Mach's idea could not be right. He himself had been convinced for about twenty years that it was right! So much for immediate recognition. One further thing, directly related to the Procrustean-bed problem, that Einstein failed to realise was that in making the gravitation field dynamical (he knew it allowed gravitational waves analogous to the electromagnetic waves on Maxwell's theory), he should have granted it a role along with matter in determining the local inertial frames of reference. It was only soon after his death that this was recognised. Einstein surely deserved not one but six or seven Nobel Prizes, but the confusion caused by his indirect attempt to implement Mach's idea is unfortunate. One can now find many differently formulated Mach's principles in the literature.

Certainty is never possible in science, but in the next part of this chapter I will present a formulation which I am confident comes closest to Mach's intuition. It matches his conviction that there must be a "unity of the All" and shifts the gaze from laws of nature found locally in inertial frames of reference to a law of the universe that explains their existence. I will present the key idea not in the spacetime terms of the general theory of relativity but rather for N point particles in Euclidean space. This will lead to a Machian—relational—form of the N-body problem that is a true unity and, as I will explain in the part of the chapter after that, points the way to its recognition at the heart of Einstein's theory.

The central thesis of this book is that the origin of time's arrows must be sought in the overall structure and behaviour of the universe. A clear hint of the desired origin is already found in the minimal model. There are arrows, with identical form, in all of its solutions precisely because the model is relational and thus has vanishing energy and angular momentum. General Newtonian solutions do often exhibit arrows but, in not satisfying the relational requirement, many lack the perfection of those in the minimal model. They do not, far from the Janus point and thus 'in the fullness of time', create the Kepler pairs that, all in one, are such precise timepieces, rods, compasses.

Newton wanted the laws of motion to match the perfection of God he saw in ideal geometrical figures. Manual craftspeople could not match the art of the divine geometer. Hence Newton's saying that "the errors are not in the art, but in the artificers". Nobody who examines the geometrical figures and proofs in Newton's *Principia* will doubt his excellence as geometer and dynamicist of the here and now. But the link to the star-studded heavens that he attempted may be a bodge job.

Timeless Best Matching

Newton's absolute space, the properties of which are replicated in any inertial frame of reference, must be distinguished from Euclidean space. In Fig. 6 the same triangle, with particles at its vertices, is shown with different positions (the central dot in each case shows the position of the centre of mass of the particles) and orientations in space (with the third dimension suppressed to permit representation on the page). Newton, who took space to be real and called it absolute, regarded the three possibilities *a*, *b*, *c* as distinct. In contrast, in Euclidean geometry they are just different representations of the same figure. In fact, since Euclidean space does not come with a notion of scale, only ratios are well defined in it. I noted earlier in the book that in the case of a triangle two internal angles define its shape; these are determined from the dimensionful lengths of the triangle's sides by forming from them two independent and dimensionless ratios. As emphasised several times, a main aim of this book is to describe the cosmos

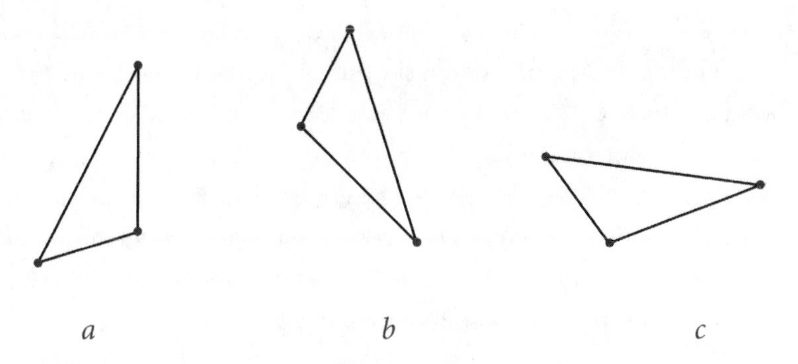

FIGURE 6. The same triangle at different positions in Newton's absolute space.

using nothing but pure numbers obtained as ratios. That can be done only by taking scale out of the picture. We will see that this turns out to be an enterprise that is subtle and with more than one surprise; it is best discussed after some preparation in which the notion of scale is retained. It will play the role of Wittgenstein's ladder—to be discarded once it has been used to climb the wall and see what is on the other side.

Disagreeing with Newton, Leibniz and Mach argued that only relative positions are observable; if the three particles are the entire universe it is a fiction to imagine there is any difference between the cases *a*, *b*, and *c*. Observers within such a universe can see only where the particles are relative to each other, not where they are relative to absolute space. In a conception of the world containing only observable things, the particles and the separations between them are all that is. It's hard to argue with that. In a renowned correspondence with Samuel Clarke, who was clearly guided by Newton, Leibniz maintained that space is not a thing, a substance, but "the order of coexisting things". By order he meant the relative distances between things. The problem, which Newton saw much more clearly than Leibniz, is when all the separations between the things in the universe change. How, under such circumstances, do you identify the same position in space at different instants of time? You must be able to do that if you want to formulate laws of motion for individual bodies. When challenged right at the end of his life on that, Leibniz had only a feeble answer. He said that "a sufficient number of bodies should not change their positions among themselves"; they would then provide a frame of reference to define the motions of the bodies that are in relative motion, both with respect to each other and to the ones at mutual relative rest. But the whole problem arises under the assumption, which Leibniz could hardly have rejected, that *all* bodies are in motion relative to each other. Leibniz's response was clearly a cop-out.

THE NEXT STEP, which is critical and which you might like to tackle after a caffeine boost, is to contrast two different ways in which something called *action* is defined. As the name suggests, it is a measure of change between two states. It is a fundamental quantity used to formulate dynamical laws, both classical and quantum mechanical, in *the principle of least action*, the generality of which is almost unlimited. It can be used to describe atoms, how an apple falls (there would be more action if it fell in a curve rather than straight to the ground), how the moon goes round the earth, and how the whole universe evolves. It involves calculating at each instant an infinitesimal action. There are two ways to do this: one Newtonian, the other Machian. They differ in the treatment of both time and space.

To begin with time, Newton's "flows uniformly without relation to any thing external" and defines both chronology (the succession of states) and duration (the length of time between the states). Leibniz countered with the claim that time is nothing but "the succession of coexisting things". Cards of a suit, say clubs, ordered from ace to king give you the idea. Mere succession, with no preassigned duration between the coexisting things in each concrete instant, is Mach's notion too. The denial of absolute duration disposes of the first of Newton's "two imposters"—absolute space and time—and prepares the ground for the *emergence of duration*. In fact, we already saw it in the creation of the ticking clock in the minimal model. In this chapter, the clock will be embedded in a comprehensive structure of emergent space and time.

The Machian treatment of the motion of matter in space cannot avoid using mass ratios; the concept of mass was one of Newton's great innovations and ever since has played a central role in dynamics, but otherwise only the notions allowed in Euclidean space are needed. They suffice to make clear what an "infinitesimal contribution" to a Machian action looks like. If my explanation works, you will have gained a good grasp of the things that determine how a succession of triangle shapes like those in Chapter 7 emerges from a relational law of the universe. Only the barest minimum of Euclidean geometry is used. Fig. 7, in which the notion of scale is retained, illustrates the two different kinds of infinitesimal action.

In the Newtonian case, three equal-mass particles in two dimensions move in the brief span d*t* of absolute time from the same initial position to different subsequent positions: one from point 1 to 1′, another from 2 to 2′, and the third from 3 to 3′. The observable changes are the same (the two solid triangles have the same side lengths, as do the two dashed triangles) but the final positions are different because the dashed triangle in *b* is rotated and translated in absolute space relative to its position in *a*. Moreover, Newtonian absolute time passes independently of whatever happens in the universe, so the changes in *a* and *b* can take longer or shorter times d*t*. For the same intrinsic change of the triangle (the change in the side lengths), the amount of Newtonian action, dA_{Newton}, is

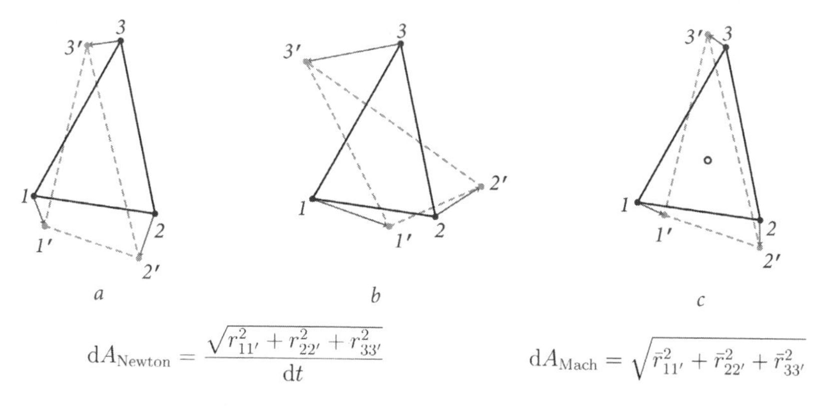

FIGURE 7. Infinitesimal actions: Newtonian (left) and Machian (right).

different because the displacements in absolute space are different and also because dt can be different in the two cases. This brings in arbitrary extrinsic aspects. Since only relative changes are observable, nothing that accounts for the difference 'can be seen'. I will say what the consequences of the two extrinsic features are after explanation of the Machian infinitesimal action, illustrated in *c*.

Everything is now intrinsic, nothing extrinsic. Besides the masses (here assumed equal) only the differences in the interparticle separations are allowed to count. The two slightly different triangles, dashed and undashed, are simply imagined initially placed relative to each other in any manner, as in *a* or *b*, but the dashed triangle is then translated and rotated relative to the undashed one into the position shown in *c*, in which the sum of the squares under the square root sign is minimized, this being indicated by the bars of the

$$\bar{r}^2_{11'}, \bar{r}^2_{22'}, \bar{r}^2_{33'}.$$

When this minimum is achieved, the two triangles are best matched: the centres of mass are brought to coincidence (at the little circle) and the rotation of the masses measured from the coincident centres of mass, weighted by the masses and taken with opposite signs for clockwise and anticlockwise displacements, is zero. Locked in the best-matched position, the two triangles can be anywhere in space without the Machian action, dA_{Mach}, being changed; its value is *background independent*. Best matching places configurations relative to each other, not relative to invisible space. Moreover, Newton's increment dt of absolute time has been eliminated; it makes no appearance in the foundations and only emerges during evolution. Note also that the value of dA_{Mach} comes out the same whichever triangle is assumed to come first in time and which is best matched relative to which. It is time-reversal symmetric.

Best matching has three effects: rotational minimisation makes the angular momentum, which otherwise could have any value, vanish; translational minimisation makes the centres of mass coincide, though without changing anything observable (Galilean relativity still holds); and the absence of dt in the Machian case enforces vanishing of the total energy, which otherwise could also have any value. Thus, Machian dynamics eliminates all traces of the arbitrary effects of absolute space and time. It is the most parsimonious dynamics possible. It cannot avoid using Newton's concept of mass but otherwise begins and ends with Euclidean geometry. In fact, its sole tool stays faithful to the fundamental principle employed in geometrical proofs, modified only by the need to allow figures to differ. Let me explain.

In Euclidean geometry, the fundamental problem that arises in proofs, for example in Pythagoras's theorem, is whether two geometrical figures—triangles are the first nontrivial examples—are *congruent* or *similar*. The figures one considers can be imagined located initially anywhere in space. To test whether they are congruent one supposes either of the figures moved and rotated wholesale through space to see if it can be brought to exact overlap—that is, laid exactly on top of the other so that they appear identical. If so, *congruence* has been established. It can be for the triangles shown in Fig. 6; they are in effect the same triangle imagined to be in three different positions. Anticipating the later discussion of scale, there can also be *similarity* of figures. In the example of triangles, this occurs despite difference of sizes provided they have the same shape. If so, they can still be brought to perfect overlap if one allows not only translations and rotations of one relative to the other but also relative scaling. This proves that the figures are similar.

Deferring to the final part of this chapter the discussion of scaling, we see already with the translations and rotations considered in Fig. 7 that in Machian dynamics considerations based on congruence cannot be taken over unchanged. In each pair of triangles in Fig. 7, the solid and dashed triangles differ in both size and shape. There is no way in which the triangles depicted at different stages of evolution could be brought to exact congruence. Nevertheless, something very like congruence is the primary tool. It is used not to establish identity but to calculate action. The two nonidentical triangles in each pair of Fig. 7 can never be brought to perfect overlap, to congruence, but in the best-matched position they are, in a well-defined sense, brought as close to it as is possible. Their incongruence is minimised. Machian dynamics is based on a *principle of least incongruence*.

So far, we have only seen how infinitesimal amounts of action are calculated when the difference between two 3-particle triangles is infinitesimal. Now we need to consider large changes such as are found in extended evolutions like those in Fig. 5 (page 105). For this, imagine the collection of all possible triangles and call it *triangleland*.* Each point in it represents the shape and size of a triangle. It is important that the triangles are not supposed to be located anywhere in either space or time. That kind of representation does not exist in advance of the dynamics but rather emerges from it. There are in fact many different geometrical representations of triangleland. For example, the x, y, z axes in a three-dimensional rectangular coordinate system could represent the three sides of the triangle. Any point for which they take positive values that satisfy the triangle inequality—that is, any one side must be less than or (when the triangle collapses to a line) equal to the sum of the other two—represents a possible triangle. The collection of all the points that satisfy the condition is an avatar of triangleland. However, just as Hindu gods can take many different forms, there are many different representations of triangleland. The important thing is that any continuous curve through any one of them represents a possible succession of triangles. As yet, that has nothing to do with location in space or time, though spatial relationships are involved in the definition of a triangle or possible figures defined by a greater number of points.

Just as there can be smooth paths through a countryside, there can be smooth paths through triangleland. Each corresponds to a succession of triangles. Select any two points in triangleland corresponding to triangles with different sizes and join them by all possible smooth curves (that is, successions of triangles that change smoothly between the two chosen points). Work your way along each curve in small steps, calculating and keeping a tally of dA_{Mach} as found in Fig. 7 for each step as you go. The sum found for the complete path is an approximation to its action. As you make the

*For readers familiar with my book *The End of Time*, this is Platonia for a three-particle universe, one for each set of given mass ratios.

steps in dA_{Mach} have ever smaller values by making the differences between the successive triangles ever more infinitesimal, you get an ever better approximation to the true action, which is obtained by going to the limit of vanishing step size. The path along which it is smallest is the path determined by the relational—Machian—law of the universe. By the way it is constructed, the universe as a whole must have vanishing energy and angular momentum, but to illustrate that geometrically we need to see how inertial frames and duration emerge from the mere succession of triangles.

This is done in Fig. 8. In it, the successive triangles, depicted as if drawn on cardboard and found by the best-matching process described in Fig. 7, are placed one on top of each other so that each next triangle is in the best-matched position relative to its predecessor. This is *horizontal stacking*. The triangles are separated in the vertical direction (*vertical stacking*) by an amount equal to the duration separating them in accordance with Newtonian motion with zero total energy (see the notes for more details). The shapes and sizes of the triangles stacked in this way create an emergent relational framework identical to that provided by Newton's absolute space and time. Within it the particles satisfy Newton's laws of motion exactly in the form they take in an inertial frame. If there are many more than three particles, exactly the same best matching

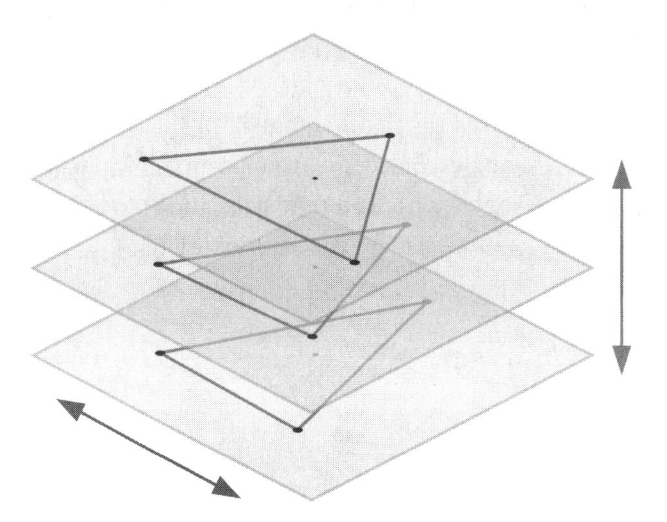

FIGURE 8. Horizontal and vertical stacking.

and stacking can be performed for them all together. This will determine an emergent Newtonian spacetime like the one in Fig. 8. Three not-too-massive particles somewhere close to each other in the middle among the totality of the particles will make only a tiny contribution to the horizontal and vertical stacking but will feel its effect as a local all-powerful inertial frame of reference.

This shows how motions can arise like those in the solar system, in which there is angular momentum in an inertial frame of reference that can be determined, without reference to the stars (which could be hidden by dense dust clouds), by earthbound observers of the relative motions of the sun and planets. The origin of the inertial frame would be a mystery for the earthlings unable to see the stars, but not for a godlike observer able to survey the whole universe and see how the greatly predominant mass of the universe determines the inertial frame with only minimal participation by the masses of the solar system.

Newton's prime argument for the existence of absolute space was the undoubted existence of centrifugal force in bodies that have angular momentum because they spin, the effect of which is manifested in centrifuges and 'wall of death' rides at fairs. Newton argued that such a force could arise only from rotation relative to some real thing. He was right about that, for to formulate the locally observed laws of nature he did need a framework, but he was wrong to reify it as absolute space. A relational law of the universe implemented through best matching creates the frame that is 'sensed' locally. It is the universe that prevents you falling to the ground in the wall of death. The implication of Fig. 8 is that space and time are not containers; both are constructed from the triangles. They are 'tied' into their successive horizontal positions by best matching and held apart vertically by the 'props' of emergent duration. Newtonian spacetime is a house of cards.

Machian General Relativity

Best matching also plays a fundamental role, though in a much more sophisticated form, in Einstein's general theory of relativity.

I will give you a brief idea of what is involved (with more details in the notes). Instead of determining how particles in Euclidean space move relative to one another, general relativity—when expressed in dynamical terms—tells you how curved three-dimensional geometry changes in time. That is, of course, very hard to visualise, but once again images in two dimensions help. Thus, suppose a ball whose surface is not spherical but undulating with dips and bulges all over. Smooth curves drawn on its surface in any way you like will intersect at definite angles. Along with the amount of curvature at different points, such information will generally suffice to identify particular points on the ball's surface and will play a role like the inter-particle separations in the particle model.

Now think how the state of the ball can change. One possibility is to suppose its size and curvature change by increasing scales in every local region by the same proportion. That will simply increase the size of the ball without changing its shape. But there is a greatly more radical way to change the shape: alter the scale at each point by a different amount while ensuring that the curves initially drawn on the surface still intersect at the same angle. That will change the local dips and bulges all over the surface but not the intersection angles. Such a transformation is said to be *conformal*—it does not change the local form (shape). This models reasonably well what happens in general relativity, though the three-dimensionality of space leads to much richer possibilities.

The 'poor-cousin' two-dimensional model, which can be readily visualised, gives some idea what they are like. Suppose once again the initial spherical surface with its local dips and bulges, and suppose it changed conformally by infinitesimal amounts that can in principle be different everywhere. How would you say which point on the transformed surface corresponds to a given point on the original surface? There is a way in which this can be done by a much more intricate form of best matching than the one used in the particle case. The broad principle is uniquely determined, but there is a range of actual implementations, one of which stands out by virtue of certain attractive features and leads to general relativity. The greater intricacy of realisation is inevitable because, compared with Euclidean space, curved surfaces and spaces have such much richer structures. The mathematician Bernhard Riemann introduced them for any number of dimensions in a famous lecture in Göttingen in 1854.* In a four-dimensional guise as spacetime, they embody the content of general relativity.

The essential feature of best matching of two slightly different curved spaces is that one can no longer imagine moving one as a totality relative to the other into the closest approximation to 'overlap'—that is, to the achievement of least incongruence. Instead one imagines doing this by trying to match up each small region in one space with every other small region in the other and determining the difference between the two complete spaces by adding up all the differences. This is not such a horrendously vast process as it might first appear. This is because the two spaces are assumed to differ only slightly. Imagine changing—as nature in fact does—the whole surface of the earth by a very small amount from its state yesterday to its state today. You know in advance there is no point in trying to match up one of the Rocky Mountains with Everest. You are never going to get a close match. In fact, you just see how each mountain today differs from the same mountain yesterday and then only in small adjacent regions. Of course, if there has been a quake overnight in California that restriction is no longer possible. In such a case you need to imagine making the comparison after a much shorter time interval—not twenty-four hours but a split second. Thanks to the power of the infinitesimal calculus that Leibniz and Newton discovered independently, that can be done.

Best matching leads to a derivation of general relativity (briefly outlined in the notes) that is quite different from the one followed by Einstein. As I noted, he did not attempt to implement Mach's ideas directly. Besides providing a different derivation, best matching

---

*Carl Friedrich Gauss, regarded by many as the greatest mathematician of all time, had proposed the topic as one from which Riemann could choose as his habilitation lecture (still the rite of passage for those who aspire to lecture at German universities). Asked after the event for his opinion of the lecture, Gauss, never one to use many words when few will do, responded: "Exceeded my expectations".

suggests, rather remarkably, that the defining feature of relativity—that there is no unique definition of simultaneity in the universe—may need to be reconsidered. This does not mean there will be changes to the outcomes of any experiments that can be made locally; the alternative derivation does not undermine the foundations on which Einstein laid his theory but rather explains, at least in part, why they exist. Indeed, conditions which generalise very directly the horizontal and vertical stacking of the particle model hold locally everywhere in spacetime and are nothing more or less than four of the ten terms in Einstein's field equations.

At the same time, there is a very great reduction, indeed huge, in the number of solutions general relativity allows as possible universes. In particular, three-dimensional space must close up on itself in the same way that the surface of a sphere does. At a stroke this enhances, literally infinitely-fold, the predictive power of Einstein's theory. The main reason for this is that general relativity, when not derived by best matching, allows space to be flat, like an infinite sheet of paper, and also allows it to be saddle-shaped with infinite extension. Current observations indicate that the universe is very close to the flat possibility but do not rule out either of the other two. For a variety of reasons to be discussed in Chapter 18 most cosmologists believe the spatial geometry of the universe is, apart from localised curvature, flat. However, some, like me, favour the closed possibility with its Machian inner unity, as did Einstein himself for at least a decade, because then, as he put it, "the series of causes of mechanical effects is closed". This was undoubtedly the main reason why, when he laid the foundations of modern cosmology in 1917, Einstein chose a spatially closed universe as his model. I should mention here a potential conflict between the possibility envisaged in this book of never-ending structure formation in the universe and the fact that, in accordance with standard cosmology, a closed universe has only a finite lifetime. Having begun with a big bang, it grows to a maximum size, starts to contract, and ends in a big crunch. I will address this issue, which is serious, at the start of the final chapter. It can at least be said that as long as the universe continues to expand, structure formation as I define it will continue.

AS INDICATED EARLIER, best matching even casts some doubt on the status of the relativity principle, in accordance with which no physical meaning attaches to simultaneity at spatially separated points. The explanation of this by means of rapidly moving trains with flashes of light emitted at opposite ends of a carriage and observed from a station platform has been the bread and butter of books for lay readers ever since Einstein published his own book *Relativity* in 1916. Let me first give a very simplified account of how simultaneity in general relativity is defined. For this, imagine spacetime to be like a loaf of bread whose long direction corresponds to the direction of time, while a slice through it corresponds to space. Such a slice is, of course, two-dimensional but is to represent space with one dimension suppressed. As Einstein first illustrated with trains, the propagation of light is critically important in relativity and limits the kinds of slices one can make through the loaf in order to get one that represents space. If units of time and distance are chosen such that the speed of light is unity, then light travels along curves in the loaf inclined at 45°. Slices and curves with inclination less than 45° are *spacelike*; curves exactly at 45° are *lightlike*; those that rise at more than 45° are *timelike*. Spacelike slices are by no means restricted to being flat. They can undulate arbitrarily provided the 45° rule is respected. All the points on any such spacelike slice can be said to be simultaneous; simultaneity can be defined in infinitely many ways in spacetime. Each possibility is called a *foliation*. In my analogy with a loaf, with one more dimension of space suppressed, a typical foliation looks like the one on the left in Fig. 9.

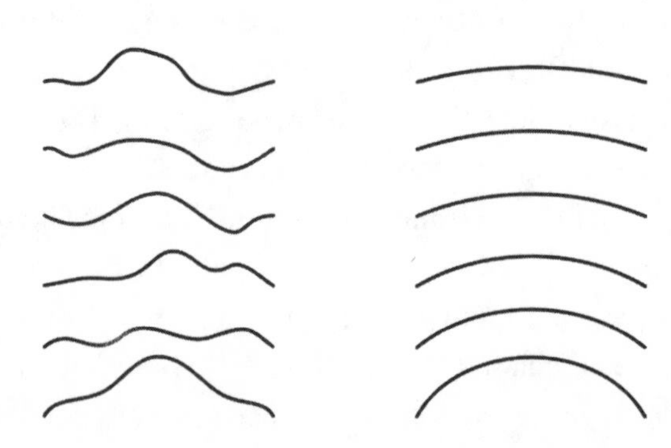

FIGURE 9. Arbitrary and CMC slicings of spacetime.

It has, however, long been recognised that there is a uniquely distinguished way of foliating space by spacelike hypersurfaces; they are called *surfaces of constant mean extrinsic curvature* (often abbreviated to CMC surfaces). Soap bubbles are the two-dimensional analogues in three-dimensional space of these three-dimensional 'surfaces' in four-dimensional spacetime. The surfaces of soap bubbles are of course not only beautiful but obviously very special, which is expressed in the fact that their extrinsic curvature, which measures their bending in the three-dimensional space in which they are embedded, is the same everywhere on their surface. Besides extrinsic curvature, surfaces are characterised by intrinsic curvature, which Gauss found is determined solely by measurements within the surface—a result so extraordinary that he gave it the name *teorema egregium* (remarkable theorem). The simplest example of it is exemplified by a flat sheet of paper, which has zero intrinsic curvature. Laid flat on a table, it does not bend in space and therefore has no extrinsic curvature. If rolled without stretching into a tube the distances between neighbouring points measured within the paper are unchanged and the intrinsic curvature remains zero. So does the extrinsic curvature, which in accordance with Gauss's calculations is the product of the nonvanishing bending of a circle round the tube multiplied by the vanishing bending of its surface along the tube—and, of course, something times nothing is zero.

I have brought intrinsic and extrinsic into the discussion here because the best-matching derivation of general relativity automatically leads to a representation of many spacetimes (when put together by horizontal and vertical stacking) in which it is CMC-foliated at least in a 'slab' on either side of a spacelike hypersurface from which one can commence evolution. Whether evolution from such a slab to a complete CMC-foliated spacetime is possible remains an open issue, to which I will return in Chapter 17. Here I will merely say that the dynamical information contained in such a slab is sufficient to determine the whole of a spacetime and, moreover, that such a CMC slab may exist in the immediate vicinity of the big bang. It will certainly be the case that within any such slice spacetime can be constructed by analogy with (but using much greater sophistication than) the horizontal and vertical stacking of Fig. 8.

There are three aspects of the best-matching derivation of general relativity and the way it leads automatically to CMC-foliated spacetimes which relate to the content of this book. First, it strengthens the case for the primacy of a law of the universe from which local laws of nature emerge. One will expect the master law to be characterised by certain architectonic features; best matching is certainly one. Second, CMC foliation will make it possible to define an entropy-like quantity for the universe associated with uniquely defined instants of time. Clausius said the entropy of the universe tends to a maximum; I hope to invert that in a well-defined sense. A notion of simultaneity will help to make both of the corresponding statements watertight. And, third, it may be noted that the absence of a good definition of simultaneity in general relativity is probably the single most important reason why it has defied for so long all attempts to cast it into a form compatible with quantum mechanics. The *teorema egregium*, a deep result in mathematics, is intimately connected with the relations that define CMC foliation. It could be a stone to kill three birds.

The Mystery and Magic of Scale

In Chapter 7, I said there is no ruler outside the universe to measure its size, so only shapes count. But then, you will ask, why not go the whole hog in Fig. 7 and best-match the sizes of the triangles? That would exploit all the freedom in Euclidean geometry and reduce everything to mass-weighted difference of shape: dynamical nirvana expressed through pure shape.

It would, but perfection can be boring; unlike Beckett's play, something does happen in a world of pure shape, but not something we would like. In fact, we couldn't exist in it. There would be none of the organic growth we observe around us. For such growth to happen, scale must play a role, and it can through ratios of size at different instants. Hubble discovered the evidence for expansion of the universe in red shifts, in ratios of the wavelengths of light emitted by atoms in distant galaxies in the past with the wavelengths of the same atoms on the earth. "It's the economy, stupid", said James Carville during the 1992 US presidential election. If it's the cosmos, it's ratios.

What would the bonus of pure-shape dynamical nirvana be? Theoretical physicists have an abiding dream: predictive power realised through a law for which the least input yields the most output. In the case of three-particle triangle shapes, it would be hard to imagine a law more powerful than one that, given two distinct triangle shapes as input, gives a unique succession of smoothly changing shapes between them as output. Adding best matching with respect to size to what is already done in Fig. 7 would create such a law. Four numbers would do the trick: two to fix the shape at one end, two at the other. In the space of possible triangle shapes, that would determine a geodesic—in a hilly landscape, the shortest path between two points is a geodesic. In fact, given the landscape and the geodesics that thread their way through it, just three numbers will identify any one of the geodesics: two determine an initial point and one a pair of back-to-back directions from it.* Mathematicians can cope with spaces with any number n of dimensions. In them, n numbers fix the initial point, $n - 1$ the initial direction. That's the kind of dynamical ideal mathematicians have in mind.

They might fall just one number short of the ideal. That would be so if two different triangle shapes *and* the ratio of their sizes suffice to generate the triangles between the two. This does not violate the ratio rule and opens up possibilities which already yield a rich harvest in the scenario in which Janus looks in opposite directions on half-universes with qualitative symmetry. Eschewing dynamical nirvana is worth the candle.

We see that already in the minimal model. Its Machian law means that absolute time, place, and orientation play no role and ensures vanishing energy and angular momentum. But the size can

---

* The meaning of 'direction' is sometimes ambiguous. From any point at which you join a footpath you can walk in two opposite directions. This is why I have sometimes used (and will continue to use when necessary to avoid confusion) the epithet 'back-to-back'. Janus should approve.

change; the ratio of the minimal model's root-mean-square length ℓ_{rms} at the Janus point to its subsequent values increases steadily. The determining data include one extra datum, the ratio of sizes, and that makes all the difference. From chaotic three-body motion at the Janus point the system evolves in either time direction into a singleton and a pair that, bit by bit, settle down to something definite and distinctive: the singleton to perfect inertial motion, the pair to a rod, clock, and compass whose direction, like love, the "star to every wand'ring bark", does not alter and is an "ever-fixed mark that looks on tempests and is never shaken". The poetry needs no apology. The one extra number in the architecture of the universe, the ratio of sizes, is what enables Shakespeare to end Sonnet 116, regarded by many as his finest, with

> If this be error and upon me prov'd,
> I never writ, nor no man ever lov'd.

If you suspect my claim is error to be "upon me prov'd", my first line of defence is the fate of the minimal model if the one extra number it allows is removed. Just as best matching with respect to rotation forces the universe to have vanishing angular momentum, best matching with respect to size forces an analogous quantity D, its *dilatational momentum*, to vanish. Whereas angular momentum measures overall rotational motion, D measures the rate at which size, expressed by the root-mean-square length ℓ_{rms}, increases. If D is zero, ℓ_{rms} must stay constant at whatever nominal value is ascribed to it. There is then no Janus point, but 'stuff' can still happen.

Indeed, one can arrange a model that, despite having no Janus point, does have gravitational forces identical to Newton's, but only if they are accompanied by further forces that keep ℓ_{rms} constant. The system can then for a while break free from chaotic three-body motion, but the singleton and pair can never properly separate. The additional non-Newtonian forces act like an elastic band that always thwarts their attempt at escape, pulling them back together. They can do nothing beautiful. Maximum predictability and anything of interest are incompatible. The next chapters will show that potentially unlimited interest is possible if the one extra datum, the ratio of sizes, is built into the architecture. It breathes life into the cosmos. Moreover, the dynamics of pure shape that the contemplated but rejected nirvana represents still serves a useful purpose in providing geodesic trajectories against which one can see the effect of the one extra freedom that must be allowed if Newtonian behaviour and structure formation are to occur: the amount by which the Newtonian curve bends away at any point from the geodesic reflects the way interest enters the world. I will call it the *creation measure* and aim to justify the name in the remainder of the book. I think it may turn out to be the most important number in the universe. It exists, as just one number, both in a Newtonian cosmos described by finitely many degrees of freedom and in an Einsteinian one that has infinitely many.

The failure of perfect predictability looks as if a bit of grit has got into the works. But it is grit in an oyster that creates a pearl. Chapter 9 will indicate how the cosmic pearl is to be recognised. Chapters 10 and 11 will explain how and why it grows.

CHAPTER 9

THE COMPLEXITY OF SHAPES

WITHOUT VARIETY THERE WOULD BE NEITHER ART NOR SCIENCE. THE UNIverse around us is fantastically varied. How might we put a number on that? Given an intuition for something, it's a good principle to seek its simplest realisation. For a model universe, we have that with mass points in space. Can we define a quantity that measures, at one instant, the extent to which they are uniformly distributed or clustered? Since no external ruler exists for the universe, the measure should not change if all inter-particle distances are increased or decreased by the same amount. The measure will then be scale-invariant, which is the mathematical way of saying it does not depend on the scale. The use of lengths cannot be avoided in the construction of such a measure. The trick is to take two different quantities that do depend on scale and form their ratio. Then the scale factors cancel and we get what we want—scale invariance.

Since particle masses play a decisive role in dynamical evolution, they also should be taken into account. But here too no arbitrary unit should play a role. Ratios of masses must be used as we already did for the root-mean-square length, but we could not do that for ℓ_{rms} precisely because a length was what we wanted. However, mass and scale are combined in the way we want in ℓ_{rms}. They are too in the Newtonian gravitational potential.

It's time for a bit more mathematics. Sorry about that; at least most of it is review. After the formulas, their significance will be illustrated with pictures. As Alice said, a book is not worth much without pictures.

Let's first recall the formula for the root-mean-square length ℓ_{rms}. For three particles of masses m_1, m_2, m_3 with distances r_{12}, r_{13}, r_{23} between them, ℓ_{rms} has the form

$$\ell_{\rm rms} = \frac{\sqrt{m_1 m_2 r_{12}^2 + m_1 m_3 r_{13}^2 + m_2 m_3 r_{23}^2}}{M},$$

where $M = m_1 + m_2 + m_3$ is the total mass. For an arbitrary number of particles, you just keep adding to the sum in the radicand the squares of the distances between all pairs of particles multiplied by the product of their masses. The dimensions of physical quantities were discussed on pages 94–95. In the expression for ℓ_{rms} the radicand has dimensions $[m^2 l^2]$, so once the root is taken and M is taken as denominator we see that ℓ_{rms} has the dimension $[l]$ of a length.

The second quantity is called the mean harmonic length, is denoted by ℓ_{mhl}, and, apart from division by M^2 to get the mass dimension right, is the negative of the inverse of the Newton potential. For three particles it is

$$\frac{1}{\ell_{\rm mhl}} = \frac{1}{M^2}\left(\frac{m_1 m_2}{r_{12}} + \frac{m_1 m_3}{r_{13}} + \frac{m_2 m_3}{r_{23}}\right) = -\frac{1}{M^2} V_{\rm New}$$

with the obvious addition of extra terms when there are more than three particles. This quantity too has the dimension $[l]$. Therefore the ratio ℓ_{rms}/ℓ_{mhl} has dimension $[0]$ and is scale-invariant. It is going to play a central role from now on and therefore needs a name. For any distribution of finitely many particles, I will call it the *shape complexity* (often abbreviated to *complexity*). Its definition is

$$\text{shape complexity} = \frac{\ell_{\rm rms}}{\ell_{\rm mhl}} = -\frac{\ell_{\rm rms} V_{\rm New}}{M^2}.$$

As a ratio, this is a pure number and therefore meets the 'gold standard': it depends only on the mass ratios of the particles and the shape of the distribution that they form. No matter which way the distribution is supposed placed or oriented in space and increased or decreased in size, as in Fig. 10, the value of the shape complexity remains the same. Mathematicians call anything with such a property an *invariant*. It's the word for something that takes on a different appearance solely because of the way it is represented, 'the way you look at it'.

Some of my readers will know that there is no universally agreed definition of complexity. Different definitions are used in different fields of study. My collaborators and I introduced the name with the definition given above in 2014. Its mathematical properties, reflected in the fact that all the variously positioned, oriented, and scaled point distributions in Fig. 10 have exactly the same value for complexity, singles it out as a fundamental quantity. Mathematicians will appreciate that: the complexity is an invariant of the similarity group. Equally important from the point of view of dynamics is that it involves not only ratios of the separations between the particles but also ratios of the masses. Newtonian dynamics is ultimately about changes in the distances between masses.

FIGURE 10. Particle distributions with the same shape and complexity.

The complexity has considerable virtues; one is that it is a sensitive measure of clustering. The obvious difference in the visual appearance of the distributions in Fig. 11 is quantitatively reflected in their complexity values. For the less uniform distribution on the left it is larger than for the one on the right. The complexity discriminates between clustered and uniform configurations because of the different ways in which its two factors vary if two or more particles move closer to each other: the root-mean-square length ℓ_{rms} changes less than the Newton potential V_{New}. In fact, in a system of many particles, a close approach of any two makes hardly any difference to ℓ_{rms} but a lot to V_{New}, which actually becomes infinite if the two particles coincide. Such behaviour has long presented problems in physics, and theoretical physicists have struggled to push them away with a device called renormalisation. However, as in the Brothers Grimm story of the frog prince, a coincidence of particles that at first seems to be a frog may turn out to be a prince when we come to consider total collisions in Chapter 16.

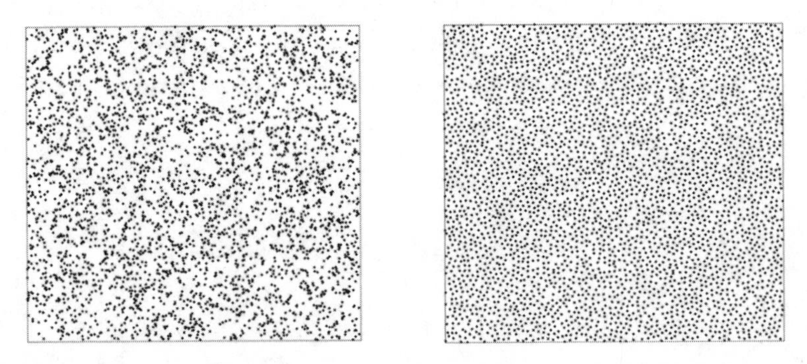

FIGURE 11. Visibly dissimilar particle distributions with different complexities.

ONE FACT ABOUT shape complexity is going to play an increasingly important role in the book. The purely mathematical desire to have a scale-invariant quantity that measures clustering of point masses leads very naturally to it as the simplest candidate. A mathematician who has never heard of gravitation but is asked to construct a quantity that measures clustering could easily come up with complexity. In fact, it puts Newtonian gravity in a new light. By itself, without multiplication by ℓ_{rms}, the Newton potential V_{New} generates forces of two kinds. Forces of the first kind pull all of the particles towards their common centre of mass, thereby changing the size of the system but not its shape. Observers within such a system, reliant solely on the determination of ratios, would have no direct means to observe the effects of such forces. In contrast, forces of the second kind do not change the system's size but do change its shape. If we reverse the sign of the complexity and regard it as a potential that generates forces, the first factor, ℓ_{rms}, kills any generation of size-changing forces. Only shape-changing forces remain. They are precisely the ones directly observable for internal observers.

This means that complexity, with its sign reversed, can be called the *shape potential*, V_{shape}. In fact, N-body specialists coined the expression in the 1990s, though for formal mathematical reasons rather than the argument I have just given. What I find interesting—it seems hitherto to have escaped notice—is that for observers within a gravitational universe, Newton's theory seems to have been constructed in such a way as to change, in the most direct way possible, the universe's most immediately observable property: its shape and thereby the variety within it. Although, apart from the minus sign (which in any case is nominal), one and the same mathematical formula expresses the complexity and the shape potential, I will, for what I think is a good reason, use whichever term is more appropriate in the given context. One dictionary definition of 'potential' is something which has the capacity to develop into something in the future. We will see that the 'something' is ever more richly structured complexity.

FIG. 12 ILLUSTRATES the powerful clustering effect of gravity. The complexity increases in either direction away from the Janus point. The plot on the left is for the three-body minimal model, its

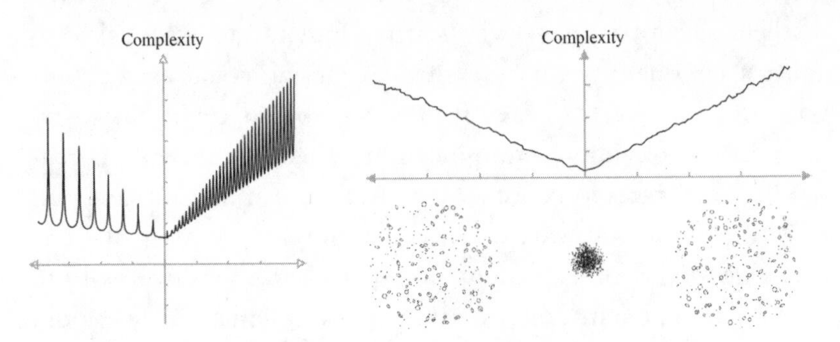

FIGURE 12. Complexity growth in 3-body and 1000-body simulations.

complexity fluctuates strongly because the Kepler pair is formed with a large eccentricity, which means that the separation between the components of the pair changes rapidly. The overall upward trend is because the size of the system, measured by the root-mean-square length ℓ_{rms}, increases steadily. This is the explanation in conventional Newtonian terms; the direct explanation is in terms of the shape, the most fundamental aspect of which is the evolution to triangles in the form of the shards with ever sharper points of Fig. 5. In the 1000-body simulation, all particles are moving relative to each other and the fluctuations are almost smoothed out. The sizes in the artistic impression of the distributions at the Janus point and after clusters have formed are nominal. Only shapes count; if magnified to the size of the outer clusters, the central swarm would be clearly much more uniform than they are. Since change in complexity is the aspect of the universe's behaviour that is most directly visible for internal observers, they must necessarily use it as the primary criterion for defining a direction of time. Observers on the two sides of the Janus point will have the indicated opposite directions of time. In arrow-of-time discussions, it has long been recognised that an initially uniform distribution like the central swarm will break up under gravity into clusters, as on the right of Fig. 12.What has not been recognised is that, provided the energy is non-negative, there must be a most uniform state in the middle of every solution with qualitatively similar behaviour on either side.

ANOTHER USEFUL PROPERTY of complexity arises because it involves only mass and distance; it can therefore be defined and used regardless of which forces—not just gravity—cause matter to clump. Cosmologists often emphasise that, on average, the matter distribution in the universe is very uniform. However, the scale on which the averaging is done is critical. The statement that the matter distribution is uniform is true if the averaging scale is taken to be greater than a few per cent of the Hubble radius, which is the distance from an observer at which the recession velocity of a galaxy equals the speed of light. Current observations suggest it is about 13.7 billion light-years.

The uniformity of the universe at this scale in its current epoch is undoubtedly a significant fact and confirms that the universe satisfies the Copernican principle: no location within it is distinguished compared with any other. However, if one looks on smaller scales, the matter distribution in the universe is very far from uniform. Expressed in terms of individual hydrogen atoms, the current estimate for the mean density of ordinary matter in the universe is one such atom in five cubic metres. For the earth, the corresponding value is more than a trillion trillion times greater. Even when the average densities of the enigmatic dark matter and dark energy are taken into account, the ratio of the density in planets and stars to the density in interstellar and intergalactic space is only reduced by a factor of about 20 and is still huge.*

---

* It is remarkable that the ordinary matter of which we, the earth, and the stars are made—and can be seen because it interacts with light—makes up less than a twentieth of the matter in the universe. About one-quarter takes the form of *dark matter*, which exists in haloes around galaxies and seems to interact with ordinary matter only through its gravitational effect. The remaining matter in the universe, somewhat more than 70 per cent of it, takes the form of *dark energy*; in accordance with Einstein's insight, mass and energy (related by $E = mc^2$) are just different names for the same thing. Dark energy manifests its existence solely through the fact, discovered in 1999, that the expansion of the universe, far from slowing down as had been universally expected, actually began to accelerate some billions of years ago.

In fact, the use of a quantity like shape complexity in the very early universe rather than the current epoch is quite interesting. The current broadly accepted view among theoretical cosmologists is that in the very earliest time after the big bang there were no particles bound in clusters. The first significant structure formation occurred when, under the action of the strong nuclear forces, nucleons were formed. They consist of pointlike particles called quarks held together by other particles called gluons. Due to the expansion of the universe, the region in which they are confined within a nucleon is very soon small compared with the distances between the nucleons. Interpreting quantum systems in classical terms has limited validity, but if we liken the first formation of nucleons to formation of Kepler pairs and clusters in the *N*-body problem then a very substantial increase in the complexity of the universe will occur at the epoch of nucleon formation.

What is more, there has, as I noted at the end of Chapter 1, been a very marked and general tendency for structure and regular motions to increase in the universe throughout its history. Ever more complex subsystems of the universe have been forming and separating themselves off from its overall expansion. They have 'self-confined' themselves. At least to a first approximation, this effect will be captured by a quantity like the complexity. Chapter 10 will provide a natural framework in which to describe the process, while Chapter 11 will show why structure formation is inevitable in an expanding universe. The proof is remarkably simple and shows why it is structure—and not disorder—that is created.

CHAPTER 10

SHAPE SPACE

And as imagination bodies forth
The forms of things unknown, the poet's pen
Turns them to shapes and gives to airy nothing
A local habitation and a name.

—WILLIAM SHAKESPEARE, *A MIDSUMMER NIGHT'S DREAM*

EVIDENCE FOR EXPANSION OF THE UNIVERSE HAS BECOME SO WIDELY accepted it is easy to suppose astronomers actually see it happening. In fact, that's a misconception, fostered in part because the expansion is often illustrated by the blowing up of a balloon. It is pointed out that spots on the surface of the balloon all move further apart from each other as the balloon expands and that this happens in such a way that ants anywhere on the surface of the balloon all see the spots receding in exactly the same way. As far as it goes, the balloon analogy is good in dispelling the idea that all the matter in the universe is expanding away from a single point. Instead, at any point in the universe observers all see essentially the same thing, which they interpret as expansion.

However, the analogy has its dangers. First, as an observational fact, astronomers do not see galaxies moving. Like Hubble, they see that the light from more distant galaxies is redder than the light from nearer galaxies. This famous red shift is expressed through ratios of wavelengths. All the evidence comes through pure numbers, nothing more and nothing less. It is all too easy to miss the fact that expansion is not directly observed. It is not as if you are looking at a balloon being blown up by a child standing in a room. In such a case, the balloon undoubtedly gets larger compared with the size of the child and the background that the room provides. These relative changes are objective but need qualification if there is no background at all. This is the situation when we consider the universe—it is, by definition, everything.

People often ask: What doesn't expand when the universe does? The answer generally given is that certain structures stabilise and no longer participate in the overall expansion. For example, atoms and molecules all strictly maintain the same size relative to each other and, less strictly, to slowly evolving celestial objects like planets, stars and galaxies. In fact, the galaxies themselves are subsystems of the universe that have separated themselves out as more or less isolated systems. What cosmologists actually establish on the largest scale is that the ratio of the typical diameter of galaxies divided by the typical distance between galaxies decreases with time.

This leads to a refinement of the expanding-balloon model that one often finds in more recent popular science books. Instead of pointlike spots marked on the surface of the balloon, one is invited to imagine rigid coins stuck onto its surface. The coins, just like atoms and molecules, all maintain the same size relative to each other but the distances between them increase. This is definitely a better analogy, but it is still misleading because it suggests a background against which you can see the balloon expanding while the coins stay the same size. But there is no such background. Located as we are within the universe, we always see any other object against the background of more distant objects in the universe. We do not see the universe itself against the background of anything else. It is all too easy to imagine we do, probably because whenever we attempt to call up, say, an image of a triangle in our mind's eye we cannot avoid endowing it with some size. I suspect this is because when we do see an actual triangle—for example, a brightly illuminated one in an otherwise totally black room—its image is projected onto our retinal receptors; the size we see is basically the ratio of the number of activated receptors to the total number in the retina. That's why we always see a shape with a size. It's just the same when we look at the stars: Orion has a definite size relative to the black vault of the sky.

IN THIS CHAPTER, as a first step to a relational representation of the universe—that is, one expressed solely in terms of ratios—I will create one for a universe of mass points. The representation will not rely on any imagined background or external ruler to define for it a size. Instead, only its possible shapes will be represented. It is also critically important this will be done in a way that takes into account the mass ratios of the particles. Newtonian dynamics arises from the interplay of geometry and mass. For each total number N of particles, the need to take into account the masses leads to a whole family of multidimensional *shape spaces*, one for each set of possible mass ratios. Each point in any one of these shape spaces represents a possible shape of the universe in a way that takes into account the mass ratios of the particles. It is fortunate that in the simplest case in which a decent minimal model exists—the relational three-body problem—shape space has only two dimensions and has a beautiful pictorial representation as a *shape sphere*.

To see that this is a possibility, let's consider how many numbers are needed to define the shape of a model universe of N point particles with fixed mass ratios. In the usual representation, three coordinates fix the position of each particle in space. That makes $3N$ numbers for the N particles. However, three numbers fix the position of the centre of mass of the particles and three more their overall orientation. These six have no place in a relational representation. Among the remaining $3N - 6$ numbers one fixes the overall size, so $3N - 7$ numbers define the shape. In the simplest nontrivial case of three particles, we need $3 \times 3 - 7 = 2$ numbers to define the shape of the triangle they form at any instant. This makes it possible to represent possible triangle shapes by 'latitude and longitude' on a shape sphere.

As determined by a particular natural formula, each point on the sphere in Fig. 13 corresponds to a distinct triangle shape, in this case for three particles with equal masses. Any two points on the sphere at the same longitude but opposite latitude correspond to mirror images of a possible triangle. For the case of equal masses shown here, the equilateral triangle and its mirror image are at the north and south poles. Points on the equator correspond to collinear configurations, in which the triangle degenerates into a straight line. Whatever the mass ratios, there are always six special points on the equator. Three, called binary coincidences, correspond to the distance between two particles becoming infinitesimal compared with the distance to the third—in the limit, the ratio of one separation to the other two becomes zero. It's always ratios that are relevant. The other three special points, distinguished by which particle is in the middle, are called *Euler configurations* after the great eighteenth-century mathematician Leonard Euler, who discovered their significance. In the equal-mass case shown here, the central particle is always exactly in the middle between the other two, but otherwise its position depends on the mass ratios. Even more interesting than these is the final possibility, which occurs when the three particles form an equilateral triangle.

The grey shading represents the values of the complexity, which has its lowest value at the equilateral triangles (whatever the particle masses) and saddles at the Euler configurations (down to either of the poles, up to the adjacent binary coincidences).* The root-mean-square length ℓ_{rms} can be used to introduce scale into the representation. For this imagine a family of nested concentric shape spheres with radii from 0 to ∞. As the system evolves in this *extended* representation, ℓ_{rms} increases or decreases while the change of shape is tracked by a moving point on a single representative shape sphere. Observers within the universe can only be aware of the changing shape of the universe and will record its history as a curve on the shape sphere, which is called the *reduced* representation.

---

*The darker crescent with midpoint at about eight o'clock in this and some further figures is solely to create an effect of illumination. The contour lines indicate equal values of the complexity.

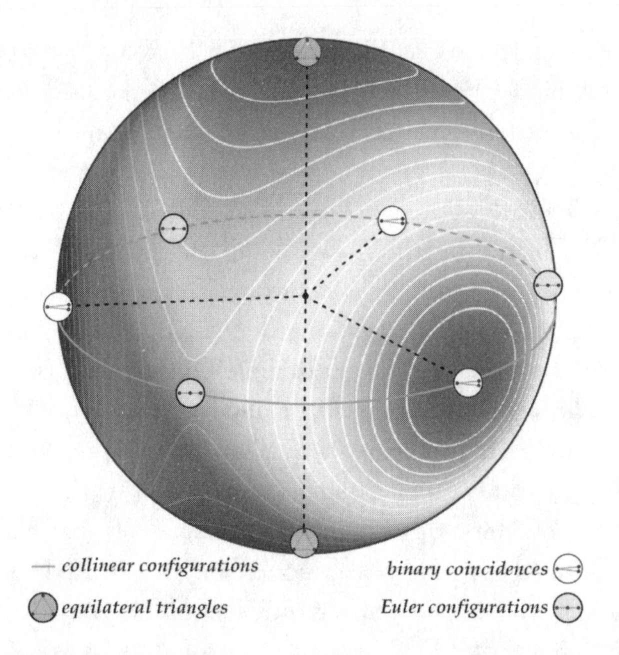

FIGURE 13. The shape sphere for three equal-mass particles.

The great thing about the shape sphere is that all the relevant aspects of three-body dynamics can be presented in a transparent manner. Moreover, several key features of relational *N*-body dynamics—for example, the growth of complexity—are already present in the three-body problem. It is lucky that all the essential features can be represented so clearly. If there are more than three particles, the situation is much more complicated. Mathematically, a multidimensional shape space with $3N - 7$ dimensions is a geometrical object just as real as the shape sphere, but I don't think even mathematicians with the very best geometrical intuition can begin to form a picture of it. Already for $N = 4$ shape space has five dimensions. It's hardly worth trying to think what that might look like. Moreover, the situation with regard to shape space is just the same as with triangleland—there are many different ways in which one and the same mathematical object can be represented. I mentioned avatars of yourself in the multiverse when talking about Boltzmann brains. The shape sphere happens to be a particularly simple and readily visualised 'avatar' of the abstract notion of all possible triangle shapes. The distinction between the representations with and without scale, the extended and reduced representations, is important whichever 'avatar' of shape space is employed.

BESIDES PROVIDING A clear picture in a nontrivial case, the shape sphere is valuable in that it highlights the critical difference between shape and size. In the extended representation, possible configurations—each defined as a shape with a size—belong to a *configuration space* that has an infinite volume because the root-mean-square length ℓ_{rms} ranges in magnitude from zero to infinity. This creates great problems for probability arguments. Suppose, for example, you want to assign a probability for the system to be in some finite region of its configuration space. If that had a finite total volume, one would have a good case for saying the probability is simply the considered region's volume divided by the total volume. This is closely analogous to probabilities when a die is thrown. It has six sides, and the probability that one particular side comes up is 1/6. But if a die had infinitely many sides, you could never win if you were to bet on any given number. As I said earlier, you wouldn't contemplate such a bet.

The situation is quite different in shape space because, as mathematicians say, it and its associated phase space are *compact* (unlike configuration spaces and the phase spaces that go with them, which are noncompact). You can get an intuition for these technical terms by supposing the shape sphere in Fig. 13 to be just one of infinitely many concentric spheres nested within each other. Ants confined to one of the spheres could never escape from it, but birds free to fly between them would live in infinite space. An individual sphere is compact; the complete collection of spheres with all radii from zero to infinity is noncompact. Similarly, a two-dimensional plane in Euclidean space is not compact, but the perimeter of a circle within it is. As already implied, the critical property of compactness opens up meaningful possibilities to introduce probabilities. The prospective prize is the ability to predict, by an argument like Boltzmann's for the attainment of thermal equilibrium, what the universe should be like near the big bang.

Significant advantages accrue from the transition to shape space. I should put to rest a worry you might have about my welcome for the finite volume of shape phase space. Wasn't Poincaré recurrence the whole problem in the tussle between Boltzmann and Zermelo? Won't the girl run out of room to tread in a compact sandpit? In a prison only boring things happen. There's no Kepler pair, born in chaos,* that becomes a clock, rod, and compass all in one. Nor, for that matter, are there girls in sandpits. The way to find them

* My occasional use of 'chaos' to describe motion near the Janus point here and later in the book should generally be understood to mean disorder and not mathematically defined dynamical chaos (often equivalently defined as sensitive dependence on initial conditions). It is, however, true that, with the usual Newtonian temporal interpretation, slight changes of the dynamical state on one side of the Janus point will in general lead to major, effectively unpredictable changes to the evolution on the other side. I'm grateful to Michael Berry for suggesting not only this clarification and the 'minimal model' title of Chapter 7 but also several other very helpful comments. Unfortunately lack of space makes it difficult to expand at any length on the alternative definitions of complexity mentioned on p. 141. Michael mentioned Kolmogorov complexity, adding that "anyway both [it and shape complexity] seem pretty weak measures of the emerging complexity you emphasise, of poets, painters, lovers, murderers, bureaucrats". That is, of course, true and I don't see any obvious connection with Kolmogorov complexity; the difficult issues in the definition of complexity in biology are discussed in the edited collection by Lineweaver, Davies, and Ruse cited in the bibliography. I do believe shape complexity, as defined on p. 140 and used in Chapter 15 as a state function to define an entropy-like quantity for the universe, is the appropriate dimensionless replacement for the dimensionful total energy, the most fundamental state function in conventional statistical mechanics.

In Chapter 17, I point out that in the case of spatially closed universes the complexity has a direct generalisation in matter-free general relativity (the Yamabe invariant), and when matter is present there may be a useful generalisation to what I call Yamabe+Matter. I think it would definitely be worth seeking further state functions that measure, for example, the prevalence of elliptically shaped subsystems within the universe and also the ratio of the total angular momentum of subsystems like galaxies and black holes to the total expansive momentum of the universe (measured in the Newtonian model by the dilatational momentum D as defined on p. 136 and again in the notes on p. 341, and also in the discussion of the Planck units on pp. 267–268). I also recommend the comments of Edwin Jaynes on pp. 294–296 about the anthropomorphic aspect of entropy definition in statistical mechanics.

In connection with biology and the second law (and Eddington's comment that "shuffling is the only thing that Nature cannot undo") Michael quoted a saying of the late Richard Gregory that is good for students to think about: "To unscramble an egg, feed scrambled egg to a chicken". On 'unshuffling', I believe this book shows that an expanding universe can and does 'unshuffle' states that near or at the Janus point are effectively shuffled.

is, once grasped, simple. As we will see in Chapter 11, the laws of motion in shape space are not those of standard statistical mechanics. They create ordered structures, not disorder. The universe of shapes may exist in a nutshell, but it created Shakespeare's verse at the head of this chapter.

CHAPTER 11

A ROLE REVERSAL

THE BEST WAY TO CONCEIVE EXPANSION OF THE UNIVERSE IS, AS I SAID, subtle. If direct observability is decisive, shape space is the true arena. Then scale plays no role. But we will now see it in a rear-guard action. It helps to work with and without scale. On the one hand, the evolution of the universe can be expressed solely in shape space. Conceptually this is significant, but the corresponding equations are somewhat complicated. In contrast, the equations in the standard Newtonian terms with scale have peerless simplicity, which makes them easier to handle. Powerful results are readily proved—for example, Liouville's theorem and Poincaré recurrence, Boltzmann's nemesis. It seems perverse to deny physical reality to scale, or at least change of scale. At the same time, there's no gainsaying the importance of direct observability. We may have to ride two horses at once, putting our weight on first one, then the other.

In fact, the problem of which horse to ride is not confined to the description of the universe. It also occurs in *gauge theories*. The great mathematician Hermann Weyl created the first one as a theory of physical lengths in 1917. He gave them the name they have because lengths are always measured relative to an arbitrary unit, or gauge. Gauge theories exist in two forms: Abelian and non-Abelian. Long after Maxwell created the laws of electrodynamics it was realised that his theory is of the first kind. Since the 1970s the more complicated non-Abelian theories have provided the basis for the study of elementary particles, with spectacular success.

Gauge theories are expressed through dynamical variables of two kinds: one called gauge, the other called physical. They are, respectively, the degrees of freedom employed in the extended and reduced representations. Scale in the *N*-body problem interpreted as a model of the universe is gauge; the ratios of particle separations, like angles, are physical. In Maxwell's theory, the electric and magnetic fields are physical; something called the vector potential is gauge. The successful use of Maxwell's equations formulated directly in terms of the fields proves they can be used without the gauge variables. However, in many cases, especially in quantum mechanics, it is well nigh impossible to make progress without the extended representation. To pass from the extended representation to the reduced representation of the *N*-body universe you need to strip from the former the three centre-of-mass variables, the three orientation variables, and above all the scale. The parallel with gauge theories is not perfect but it is good enough. In fact, a theory in which the angular momentum must be exactly zero is a non-Abelian gauge theory. The gauge degrees of freedom in this case are related to a supposed orientation of the complete system, but when the angular momentum is zero no particular orientation can be identified.

There is much discussion among theoretical physicists about the difference between gauge and physical degrees of freedom. Many would prefer to work solely with the physical ones, but little progress has been made in that direction and I'm not sure it's the right way to go. The striking thing that I see in gauge degrees of freedom is that they not only simplify the equations but also seem to help to explain the behaviour observed in the physical degrees of freedom. I am going to give an example now. It will explain the 'role reversal'.

The suggestion that time's arrows may have a purely dynamical explanation rests on the hypothesis that the law which governs the universe dictates the presence of a Janus point in all solutions. Then they will all divide into two qualitatively similar half-solutions or, when we come to the situation in which the size of the universe is zero at the Janus point, two 'bundles' of half-solutions; in both cases the time-reversal symmetry of the law which generates them will be respected. Chapter 9 showed that in the considered particle models the growth of complexity in both halves of all solutions defines dynamically dictated arrows of time. They have nothing to do with statistical effects made possible by the existence of many as opposed to few particles. The three-particle minimal model is already paradigmatic and exhibits the arrows with especial clarity.

This is a start, but it is far from a complete solution of the problem of time's arrow. Most of the remainder of the book will be taken up with what might be called supporting evidence and exploration of some of the implications of such a solution. I believe that the overall picture which emerges is persuasive, but I am the first to recognise that much work remains to be done. In fact, one important issue—the role of quantum mechanics—can hardly be addressed because, as I already said, no proper theory of a quantum universe exists. But William Thomson, with aim as sure as William Tell's when he shot an apple from his son's head, shot time's arrow not into an apple but into the very heart of physics in his dissipation-of-mechanical-energy paper of 1852—decades before the discovery of quantum mechanics. Classical models may provide significant clues.

The best clue that I know relies on Liouville's theorem; you will recall that, forced to act in a bounded phase space, it led to the recurrence theorem and Boltzmann's undoing. A patch of oil on the surface of a stream illustrates the Liouvillean half of Poincaré's proof. The stream is phase space, and the patch consists of innumerable tiny spots of oil that stand for independently evolving microstates of the system; each traces out an individual microhistory. The stream's flow distorts the patch, forcing it perhaps to grow octopuslike tentacles, though without ever detaching them. In fact, it is only for a system with a single degree of freedom that phase space has two dimensions like the surface of a stream. Realistically we have a highly multidimensional phase space and the 'patch of oil' is a region within it that occupies a certain volume.

It was not Liouville's theorem by itself which created the difficulty for Boltzmann but the extra assumption that phase space is bounded. The sandals in a sandpit illustrate the consequence. Phase space becomes a prison in which a Nietzschean eternal-recurrence nightmare gives sleepers no respite. But the girl can walk in an infinite sandpit, and streams need not, like the river Jordan, end in the Dead Sea. They can and often do reach "great Neptune's ocean". In mathematical rather than poetical terms let's consider the implication, which is easy to grasp.

It is related to something I could have mentioned earlier when contrasting the extended and reduced N-body representations, but it fits better here. If one uses three Cartesian coordinates to describe the position of a single particle in absolute space, there is nothing that distinguishes any one such coordinate from the others. Absolute space is featureless; with only one or two particles the deception it practises is hard to recognise. But let there be N particles, with N being three or more; then there are $3 \times N = 3N$ coordinates. Relationally they are far from all on the same footing. Three locate the centre of mass, three define the orientation, and one defines the scale; the remaining $3N - 7$ define the shape of the particle configuration. The lack of distinction between the Cartesian variables is an artifact of the Newtonian representation, which erects a 'scaffolding' over the shape, obscuring the true core of dynamics from the incurious view. The $3 + 3$ variables that describe the overall position and orientation have a quite different nature from the shape degrees of freedom, which for their part have nothing in common with the scale variable. Let's consider how the differences manifest themselves in Liouville's theorem.

It is formulated in the N-body extended representation with $3N$ position coordinates and as many matching momenta. The relevant phase-space volume is a product formed pairwise from each position coordinate and its momentum partner. In the relational N-body problem used to model the universe, the three position coordinates and three orientation coordinates play no role; there remain only the shape coordinates, their momenta, and, crucially, the scale and its momentum. Liouville's theorem still applies to this slimmed-down system. The phase-space volume, which must remain constant, is a product of the shape and scale parts. The critical fact now is that in general the scale part will increase in the direction away from the Janus point simply because the particles, being unconfined and in general having different velocities, are bound to spread out in space. That must increase their root-mean-square length ℓ_{rms}, which is the measure of the system's size. Since the total Liouville volume must remain constant, the shape part must decrease. In turn, this means that the system, whose micro-history cannot leave a volume whose size in shape phase space is steadily decreasing, must be forced toward certain special shapes and motions. Fig. 14 illustrates the effect.

In it, the paths traced in space by the three equal-mass particles for the solution of the minimal model in Chapter 7 are shown on the left, which corresponds to the extended representation. The corresponding solution curve in the physical representation on the shape sphere is shown on the right. The small cross in the dotted part of the curve (on the 'back' of the shape sphere) marks the Janus point. No matter where it lies, each half of the evolution curve that emanates from it is 'dragged' in a never-ending spiral to one of the three special points on the shape sphere that, in the

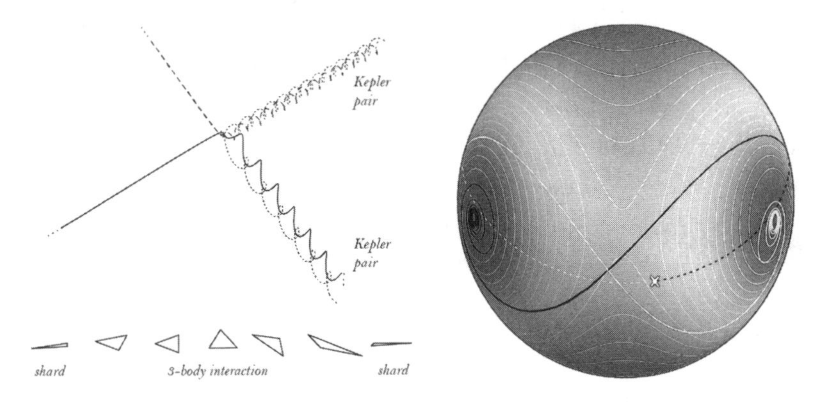

FIGURE 14. Three-body dynamics in space and on the shape sphere.

extended representation, correspond to the singleton being infinitely far from the Kepler pair. At the same time, the successive three-body triangle shapes (shown on the left) become ever more acute 'shards'. Moreover, the change of shape, which is irregular at the Janus point, becomes increasingly repetitive as the two particles of the pair circle each other in a more and more regular fashion while the singleton moves away from them in an ever straighter line. The shape changes in a very special way. The technical language used to describe an effect of this kind is that the system is forced to *attractors*. This is what Liouville's theorem imposes in all Janus-point systems, which by definition are unconfined and satisfy a law which allows them to expand without limit (as current evidence suggests is the case for the universe; it has certainly been expanding since the big bang). The corollary is that the shapes and the manner in which they change become more and more special.

If we now remember that observers within the universe only have direct access to the shape degrees of freedom, it follows from the necessary existence of the attractor effect that pervasive arrows of time must be observed throughout the universe as a consequence of its expansion on either side of its Janus point. This conclusion agrees with the long-held suspicion—first mooted, as I said earlier, by Thomas Gold in 1962—that expansion of the universe is the master arrow behind all of the other arrows of time, but to the best of my knowledge the suggestion that Liouville's theorem provides a direct causal mechanism for the creation of time's arrows is new. It is from my collaborator Tim Koslowski and came out of ongoing discussions between him, Flavio Mercati, and myself in December 2014. Other researchers had somewhat earlier noted the importance of Liouville's theorem in an expanding universe in connection with cosmological inflation, which will be discussed in Chapter 18; prominent among these was David Sloan, who began to collaborate with us in the autumn of 2015.

If the decisive effect of Liouville's theorem is confirmed, it will be an ironic role reversal. It was the very same theorem, the essential prerequisite for validity of the recurrence theorem, that thwarted Boltzmann's attempts to give a purely dynamical explanation of entropy growth and the second law; he had to include a critical statistical element. Once we lift the assumption that the universe is 'in a box' (the other necessary condition for the recurrence theorem to hold), Liouville's theorem ceases to play a spoiling role and instead becomes the prime enabler. It dictates the presence of arrows of time pointing in opposite directions from the Janus point in all solutions. This general conclusion does not yet guarantee that all the known arrows of time will be enforced and have their observed form. The minimal model so far used does broadly match what is observed in the universe—namely, a tendency for the matter in it to evolve from a rather chaotic but basically uniform state into one that is much clumpier and, above all, has positions and motions that are strongly correlated and highly structured. But that falls a long way short of predicting the kind of clumping that is actually observed. That and the other main arrows of time remain to be discussed later in the book.

WITH THE BENEFIT of these insights, let us consider once more the relative merits of the extended and reduced (physical) representations. To what extent does each allow us to understand what happens? Does one or the other supply a better intuitive explanation of observations?

In the representation with scale the universe expands. Then Liouville's theorem and the action of gravity immediately explain the attractors on shape space. This suggests that 'below the surface' of the directly observable, an invisible reality—scale, or at least the ratio of scales—plays a decisive role. David Sloan thinks otherwise. Friction is a phenomenon with which we are all familiar. Indeed, for two millennia everyone believed with Aristotle that force was needed to maintain motion. Newton changed all that with his first law; according to it the natural tendency of a body is to remain at rest or move in a straight line with uniform speed. Force does not maintain that state but changes it. What David noted, and Flavio too a couple of years earlier without the connection to Liouville's theorem, is that the universe 'takes paths' in shape space which resemble those of bodies subject to friction.

The analogy with friction is apt; imagine a smooth surface that is flat except for a circular bowl-shaped depression. If a ball rolls without friction across the surface, it will follow a straight line until it encounters the depression, where it will be deflected, and it will speed up under the influence of gravity, which pulls it down toward the bottom of the bowl. However, energy conservation ensures that the ball emerges from the bowl at the same speed and merely with a changed direction. If instead the table is sticky, friction will slow the ball down; either it will still emerge from the depression but deflected through a greater angle or, if the friction is stronger, it will spiral round and come to rest at the bottom of the depression. If our only interest is in the positions through which the ball moves and we have no concern about its speed or when it passes particular points, we could apply dye to the ball so that it leaves a track. If the friction is strong enough, the track will enter the depression, circle around it a few times, and end at the bottom.

If you now look again at Fig. 14 and interpret the darkness level as a measure of the shape potential's depth, you will see that the universe makes tracks in its shape space very like the ball on the sticky table, but with two differences. One is trivial: the ball's track starts at some initial point in a straight line and terminates, after finitely many turns, in a single spiral. There aren't two, one at each end. In contrast, the universe sports two twirls. Much more significant is that, despite getting ever smaller, the turns are infinite in number. Normal friction brings motion to a full stop. The effect here is like that of friction with a strength which weakens at precisely the rate that allows motion to continue forever. The universe has a history that never ends. Moreover, the attractors that it strives to reach but never attains are very special. We know what they are in the minimal model, but that's based solely on Newtonian gravity. What shape the strived-for but never attained attractors take in a universe governed by general relativity, nongravitational forces, and quantum mechanics is another matter.

This I do believe: we need to look for them in shape space. It is the true arena. Two things speak for it. First, not one iota of essential information is lost when, as in Fig. 14, a solution created using Newton's equations is projected into shape space. That is what Flavio did to create the figure. Second, shape space is spare; it contains nothing that is redundant. The elimination of scale enables one to see history as it is. It was on the road to Damascus that the scales fell from the eyes of St Paul. Passing from the plural to the singular, Tim said that one can only see the true shape of the universe when the scale falls from your inner eye. Newton's equations do have a beautiful simplicity but, as we will see, it veils real beauty.

CHAPTER 12

THIS MOST BEAUTIFUL SYSTEM

IF THE THREE-BODY PROBLEM GAVE NEWTON HEADACHES, THE *N*-BODY PROBlem might have induced migraines. Rigorous results are almost impossible to come by. Luckily, some qualitative conclusions of much interest can be drawn. They are especially striking if, as in a relational universe, the energy and angular momentum vanish. In the three-body case that is what ensures, in every solution, the emergence of a rod, clock, and compass all in one as the Kepler pair stabilises. What can we expect with a multitude of particles?

The typical behaviour is broadly analogous. The system breaks up into subsystems. In each, energy and angular momentum are conserved with ever better accuracy. Moreover, any two subsystems move apart from each other with a speed that is proportional to the distance between them. This is exactly what Hubble found for intergalactic separations and proved expansion of the universe. This was and still is regarded as a remarkable consequence of Einstein's general theory of relativity, though if the *N*-body clusters are taken to model galaxies, similar behaviour is inherent in Newton's theory.

There's a rich and remarkable story here, of structure emerging out of chaos. Not quite the chaos of the ancient Greek myth, in which 'chaos' refers to a vast chasm or void. That's also what the Book of Genesis says: "The earth was without form, and void".

The chaos I mean is disorder. In Chapter 14 I give probabilistic arguments for the form it might take. For the purposes of this chapter let us assume that at or near the Janus point the system of particles resembles a more or less uniform swarm of bees that fly in all possible directions with random speeds. The three-particle minimal model of Chapter 7 gives some support for this assumption—the only truly disordered and chaotic state in its history is at the Janus point. What can be said with certainty is that at that point the system has vanishing energy and angular momentum, as it does throughout its history, and in addition—uniquely at that point—no overall change of size.

Since at any given instant it is never meaningful to ascribe a size to the universe, in saying what it 'looks like' at the Janus point we should only consider its shape there, though we can give it a nominal size. The law that governs its evolution will then give a current size relative to that nominal one. There is an important caveat if the size at the Janus point is zero, so that in Fig. 2 (page 96) the U-shaped curve on the right, which touches the time axis, replaces the one on the left, which does not. Then a size must be specified at some other point. Leaving for later this exceptional case, which is very interesting as a model of the big bang, let us now start on the promised story.

It's more or less what happens in the three-body universe, writ large. First, the initially uniform state breaks up into clusters. At the same time, the system's size (relative to its nominal value at the Janus point) increases in the way Lagrange found in 1772. Just as the singleton and Kepler pair separate in the minimal model, so do the clusters. They become progressively more dynamically isolated from each other and their centres of mass approximate ever better inertial motion. As their speeds stabilise, the distance of each cluster from the complete system's centre of mass increases in proportion to the time. In time, all clusters are moving inertially with different speeds away from a single point, the common centre of mass. More significantly, the distance between any two clusters increases, with ever better accuracy, at a rate proportional to the distance between them. This is the Hubble law already mentioned.

There will be different numbers of particles in the individual clusters, most of which will exist only transiently, losing particles until typically only singletons and Kepler pairs remain. That this breakup happens is significant; its form is of little concern.

TO ADD VERISIMILITUDE to the story, imagine beings at the particles within a cluster who can make observations and communicate with each other. This will enable them to determine how the separations between its particles change. What they and observers in other clusters find adds to the wonder of the ballerina's pirouette. As it happens, it was Lagrange himself, in his 1772 essay, who in a mathematical tour de force showed that from observations made entirely within a cluster—that of the earth, moon, and sun in his case—it is possible first to establish that the cluster is becoming perfectly isolated and then to identify a counterpart of the vertical line through the ballerina's body about which she pirouettes. That's what Newton's laws can tell you when expressed not relative to absolute space but in terms of the inter-particle distances. It's not that if one were looking at the constantly moving particles, the line would stand out like the ballerina's, which you see framed by the stage and proscenium. It took a mathematician of Lagrange's calibre to see it. At any instant, it has a definite position within the relative configuration of the particles. Its existence can be established within a cluster of three or more particles without reference to the rest of the universe, which could be obscured by thick dust clouds. There is just one exceptional case in which there is no such line: when the cluster's angular momentum is zero, no direction is distinguished, but there is virtually no chance of this being the case. So far we have considered what can be observed within an individual cloud-surrounded cluster as it becomes progressively more dynamically isolated. By itself, the effect is already striking; if the clouds around it evaporate to reveal the whole universe, the vista is breathtaking. In the original state at the Janus point, no spatial direction is singled out for the universe as a whole because its angular momentum, like its energy, is zero. No directions are distinguished anywhere because isolated clusters have not yet started to form. But as soon as they do the motions within each of them start to acquire the character of any isolated Newtonian system: energy, momentum, and angular momentum for them begin to be defined and conserved with increasing accuracy. Well-isolated Kepler pairs form and become good rods, clocks, and compasses. In fact, all the more or less isolated systems play similar though less clearly manifested roles.

Moreover, with increasing distance from the Janus point, all the compass directions defined by the emergent clusters settle down into fixed alignments relative to each other that are maintained until the clusters break up through the loss of particles or encounter other systems (what happens then is one topic of Chapter 19). The same is true, especially for the Kepler pairs, of the time they tell and the distances they define. This is not to say that the clock periods and rod lengths are the same but rather that their ratios remain unchanged. Metrology, the science of measurement, invariably needs actual bodies or materials to provide reference units—for example, the rotation period of the earth to define time in the metric system, or quantum transitions of caesium atoms in the modern atomic clocks. In the increasingly ordered universe that emerges with increasing distance from the Janus point, our imagined observers witness the birth of metrology. Restricted to observations they can make in their current epoch, they will find a miraculously arranged universe and wonder how it could ever have arisen. That is what Newton asked in awe as he contemplated the solar system. In *The System of the World*, Book 3 of his *Principia* and written before Uranus and Neptune were discovered, he said:

> The six primary planets are revolved about the sun in circles concentric with the sun, and with motions directed towards the same parts, and almost in the same plane. Ten moons are revolved about the Earth, Jupiter, and Saturn, in circles concentric with them, with the same direction of motion, and nearly in the planes of the orbits of those planets; but it is not to be conceived that mere mechanical causes could give birth to so many regular motions. . . . This most beautiful system of the sun, planets and comets, could only proceed from the counsel and dominion of an intelligent and powerful Being.

Newton underestimated the power of his own laws. He was unable to conceive that the order within the solar system could have come about through "mere mechanical causes". He also thought that those same mechanical causes would, through the forces that the planets exert on each other, lead to the solar system becoming disordered, which would require God to intervene every now and then to restore order. Leibniz mocked Newton for supposing the creator to be such a poor clockmaker that repair interventions would be necessary. God must surely be a master craftsman. Modern science dispenses with God, replacing him with laws of extraordinary power. I will come back to that later. Here I will merely say that Newton's laws have long been known to ensure the solar system's survival as now constituted for a very long time. Moreover, together with more recently discovered laws, they go a good way to explaining not only how a solar system seemingly subject to "the counsel and dominion of an intelligent and powerful Being" can come into existence but also much of the order observed in the wider universe.

SOME THOUGHTS ABOUT the order will not be out of place. It is often said that Newton discovered a clockwork universe, with its image of remorseless cogs turning forever in dreary repetition. It's why many artists recoil from science. William Blake's visceral dislike of Newton and all his works is well known. Keats complained that "all charms fly at the mere touch of cold philosophy", which will "conquer all mysteries by rule and line" and "unweave a rainbow". In *Unweaving the Rainbow*, Richard Dawkins sought, with his gifts as a writer, to counter Keats's complaint by showing just how wonderful the innumerable discoveries of science are. I fully concur with that but wonder if his efforts will only increase the concern. For everything that Dawkins writes points in the same direction: the laws of science are all-powerful. In a paragraph or two I cannot do better than Dawkins, but my parable of the birth of metrology does challenge the clockwork universe. How and why did the concept take hold?

The main reason is obvious. The solar system does, at least over an astronomically long time span, resemble a clock—indeed several, since each planet keeps good time. There is also the role of history: Newton's was a world intoxicated by discovery and invention. Fourteen years before the *Principia* appeared in 1687, Europe had acclaimed Christiaan Huygens's *Horologium oscillatorum* on the theory of clocks. Huygens himself, in 1656, had been responsible for the invention and construction of the first genuinely accurate human-made timekeeper, the pendulum clock.

It's hardly surprising that Newton and his contemporaries saw his discoveries as evidence for the divine creation of a wonderful clockwork universe. But if the picture of a relational universe presented in this chapter, with some supporting arguments yet to come, is broadly correct, there is much more to Newton's discoveries than mere clockwork. His laws imply cosmic history hidden by the contingent stability of the solar system. Without its stabilising negative energy, there would be qualitative change and evidence of historical evolution. The first unrecognised hint of that came with Lagrange's essay in 1772. It took the discovery of the second law of thermodynamics eighty years later to bring history firmly into physics. But it could have been seen in the *N*-body problem. Its equations do not merely enforce the repetitive behaviour of clocks; they mandate the clocks' creation in the first place. Only then do they allow the clocks to tell the time. Mindless clockwork, no less than heat death, is indeed depressing. But everyone likes a good story. There does seem to be more than a little truth in Muriel Rukeyser's claim that "the Universe is made of stories, not of atoms".

In fact, unpacking a Russian doll comes to mind. There are multiple stories within one overarching story. The theme that unifies them is the effect of expansion on a universe that is, and rests forever, not in thermal equilibrium but in a state of vanishing energy and angular momentum. Compared with everything else that can be observed far from the Janus point, this condition is sui generis—in a class by itself, unique. It is 'especially special'. Why? For two reasons. First, it alone is exact. The various alignments between clusters and the conservation of energy and angular momentum within each of them only become true with better accuracy, never exactly so. Second, these facts are expressed through ordinary numbers:

the values of energies, angular momenta, et cetera. In contrast, the energy and angular momentum of the universe are exactly zero. That's special. Among all numbers, zero is unique. In his beautiful little book *Spacetime Structure*, Erwin Schrödinger says

> A great many of our propositions and statements in mathematics take the form of an equation. The essential enunciation of an equation is always this: that a certain number is zero. Zero is the only number with a charter, a sort of royal privilege. While with any other number any of the elementary operations may be executed, it is prohibited to *divide* by zero—just as, for example, in many houses of parliament *any* subject may be discussed, only the person of the sovereign is excluded.

The passage continues with more in the same vein. Schrödinger's point is that an equation always involves the vanishing of some number.* An equation can express a more or less profound truth. Vanishing of its energy and angular momentum expresses the unity of the universe. Subject always to that unity, expansion can work its magic, creating in great abundance, through the attractor mechanism, increasingly isolated subsystems. They too have their own unity; its incomplete perfection, which we see all around us, is a hint of the higher whole.

Where do we find it and what does it look like? During the composition of his Fifth Symphony, Sibelius wrote in his diary: "It is as if God Almighty had thrown down pieces of a mosaic for heaven's floor and asked me to find out what was the original pattern". Closer to the concerns of astronomers but still relating to the heavens, Copernicus grappled for years with the lack of order in the planets' motions. In his preface to *De Revolutionibus*, published in 1543, he said that the principal failing of the astronomers who

---

*The vanishing of a number has in fact dictated the course of this book since the introduction of the Janus point; in Fig. 2 (page 96) the curve of the square of ℓ_{rms} at the Janus point is flat—its slope is zero in both cases. Moreover, congruence exists when there is zero difference between two figures. Best matching is based on the principle of minimising incongruence.

had preceded him was that they could not deduce "the structure of the universe and the true symmetry of its parts". He continued: "On the contrary, their experience was just like someone taking from various places hands, feet, a head, and other pieces, very well depicted, it may be, but not for the representation of a single person; since these fragments would not belong to one another at all, a monster rather than a man would be put together from them".

When first Galileo, followed by Huygens and then Newton, found one result after another in dynamics, their results could be called pieces of mosaic; they were certainly well depicted. But Galileo's rolling balls and parabolic motions of cannonballs were described relative to the solid earth, as was Newton's falling apple. Huygens discovered the laws of elastic collision by applying Galileo's relativity principle to collisions that took place on a boat on a canal in Holland while being simultaneously observed from the bank. Of the boat he simply said it moves uniformly "relative to other bodies we assume to be at rest". But where and what are those other bodies? Newton claimed to have defined the true framework in the form of absolute space and time, but he failed to show how absolute motions could be obtained from observed relative motions. Two centuries later the stopgap of inertial frames was introduced but without relation to the universe at large. Each motion described using an inertial frame is a piece of mosaic, but not fitted into the structure of the universe.

As far as the Newtonian dynamics of finitely many mass points in Euclidean space is concerned, I suggest shape space is the true framework. All possible arrangements of the particles can be represented adequately and economically in it: once and once only as unique shapes, not multiply as in the Newtonian manner of the three-particle triangles of Fig. 6. The history of the universe is a path that winds its way through shape space. Everything that is essential is kept; everything that is not is eliminated. That includes not only scale but also time. History is not a spot that advances along the path as external time passes; it is the path.

CHAPTER 13

TWO KINDS OF DYNAMICAL LAWS

EMMY NOETHER IS RECKONED BY MANY TO BE THE MOST IMPORTANT WOMAN in mathematics. She proved a theorem said to be "certainly one of the most important mathematical theorems ever proved in guiding the development of modern physics, possibly on a par with the Pythagorean theorem". Noether was Jewish. In Germany after Hitler came to power in 1933 she was refused the right to continue teaching at Göttingen and had to move to Bryn Mawr College in the United States, where, in a brief period before her early death, she continued to do brilliant work, developing a new branch of mathematics quite unrelated to her theorem. One fruit of this work, which she did not live to see, is the proof of the surprising fact that on the surface of the earth there always exists at least one pair of antipodal points at which the pressure and temperature are exactly the same.

Einstein's general theory of relativity was the stimulus to Noether's theorem. She had been asked by the great mathematician David Hilbert to study the status of conservation laws in the general framework of physical theories that he had been developing at Göttingen. A visit to that famous university by Einstein in the summer of 1915 led to a race between him and Hilbert to find the definitive equations of general relativity. The two men found the equations more or less in a dead heat, which initially resulted in some tension between the two. There can, of course, be no doubt that virtually all the credit for the discovery of the theory is due to Einstein; it was he who identified the framework in which it was created and what it should achieve. Those conditions dictated the most appropriate equations that would meet the requirements. This was not apparent to Einstein as he struggled to find them, but with hindsight it was inevitable and only a matter of time—a few months, in fact—before the equations would be found.

Hilbert then began to consider the implications of the equations and the general framework in which they had been found for conservation laws, and especially for the principle of energy conservation, which Thomson in particular had made the monument of nineteenth-century physics. Noether attacked this issue in the framework of variational principles used to solve optimisation problems. Specific problems of this kind come in a great variety of forms. One of the most famous is Dido's problem; the legend of the founding of Carthage has it that when she came to North Africa in search of a home for her people she was granted the amount of land she could enclose between two points on the Mediterranean shore and a cowhide cut into the longest possible strip and laid on the land between them. The search among possibilities involved two elements simultaneously—the choice of the two points on the shore and the way the strip should be laid on the land between them. The two conditions together would maximise the area of the 'bite' she could take. Another well-known variational problem, that of geodesics, already appeared in Chapter 8. In it, all routes between two fixed points are 'explored' to determine their lengths. The shortest route is the geodesic.

Among all variational problems some are special in being *invariant*. 'Invariant variational' sounds like an oxymoron, but it isn't: the resolution of the apparent paradox is that the quantity which is being varied has the property of remaining unchanged when expressed in different ways. I won't attempt a formal definition, but in a moment I'll give you an idea what that means. All I'll say at the moment is that invariance is a defining property of the equations that Einstein and Hilbert were competing to find. In fact, invariance directly guided Hilbert in his discovery of the equations; Einstein found them by an inspired combination of other ideas.

NOETHER'S DECISIVE INSIGHT was that there are two broad classes of invariant variational principles, each with its own implications and different kinds of invariance. Her theorem consists famously of two parts that deal with the two cases. It's a challenge to characterise them for the lay reader, but I'll have a try. The first is a bit easier and prepares the way for the second.

Suppose in Dido's problem the Mediterranean shore were dead straight and Africa flat as a pancake. Then, for a given length of cowhide, Dido's 'bite' into Africa would clearly have the same area anywhere along the coastline. Key here is the uniform framework—the straight coastline and the flat continent, within which the cowhide can be laid out in different ways. It is through the uniformity that invariance comes in. The conditions are the same everywhere. Note also that the 'rules of the game', Dido's task, stay the same: maximise the enclosed land area. If Dido worked her way west from where Alexandria now is to see if one location gave her more land than another, she would find the same area at all points. In physics quantities whose values do not change in time are, as you know, said to be conserved.

In the dynamical problems studied prior to Einstein's great theory, a uniform framework was taken for granted: time passed like my image of Dido's rectilinear westward progress; atoms moved in uniform absolute space; Newton's laws were fixed once and for all. Moreover, these 'rules of the game' could be formulated succinctly as the principle of least action that I used in the best-matched form of Chapter 8.

What Noether's part I showed was that in all variational problems in which there is a uniform background of any kind (not necessarily of space and time) certain quantities will always be conserved in going from one part of the background to another. The most fundamental conservation law—that of energy—arises because, as formulated mathematically, there is a uniform temporal background analogous to a straight shore in Dido's problem—it's the spatialised timeline that Bergson so disliked. Noether showed that dynamical laws derived from the principle of least action will lead to energy conservation if the form of the action does not change in time (treated as external). Many scientists find this a great and reassuring result—unlike the fate of all material things, the jewel in the crown of dynamics will not lose its shine.

Of course, individual examples later subsumed in the general principle of energy conservation had long been known, and the universal formulation of the principle of energy conservation following the discovery of the laws of thermodynamics in the middle of the nineteenth century had made an immense impression. But when Hilbert turned to Noether with his request he could hardly have anticipated she would add to the story such a twist: the utterly simple explanation of energy conservation is that the laws of nature do not change in time. Laws, energy, and time are linked by a most intimate bond. However, there is the caveat, seldom flagged, that time is taken to flow without regard to what happens in the world.

If that were not enough, what Noether established in part I of her theorem was much more general and far-reaching than just the origin of energy conservation; she explained, within the encompassing framework of variational principles, how all conservation laws arise. Uniformity, either assumed in the mathematical description or effectively present in the field of application, is always behind them. Energy is conserved because time is uniform; momentum is conserved because space is the same everywhere—it is homogeneous; angular momentum is conserved because space looks the same in every direction—it is isotropic. Part I of Noether's theorem is a mathematician's delight: a simple unifying result that underlies a host of phenomena common in kind but different in detail. But it all depends on a presumed external framework. (Noether did not say that in her paper; she simply pointed out the mathematical consequences of mathematical assumptions, one of which was an independent variable that could be time or position in space.)

IF PART I of Noether's theorem gives conserved quantities, part II takes away equations of motion. In Newtonian dynamics, each degree of freedom has its own equation of motion. For example, each coordinate x, y, z of a particle in an external force field satisfies an equation that relates the acceleration in the coordinate direction to the force in the same direction. In contrast, in systems subject to part II of the theorem, not all the equations of motion are independent. Some are consequences of others. Moreover, the situation is democratic. If, say, m equations are consequences of n others, the total $m + n$ equations can always be split in such a way that m arbitrarily chosen equations are consequences of the remining n.

Part II is significant because it tells us what is real in the world. Once again the relational three-body problem is paradigmatic. In the Newtonian representation, each particle has three coordinates, so there are nine in total. For Newton, each of these is real because absolute space and position in it are real. But in the relational conceptual world only two independent ratios of the three inter-particle separations are real. If you take away an imagined uniform background, things can only change relative to each other, not also relative to a nonexistent background. This is reflected in the mathematics. There are not as many independent dynamical equations as you would expect if, taking Newton at his word, you instinctively think the world's events unfold in absolute space and time. What happens in this case is that there are far fewer possible histories in the three-body problem, expressed as successions of triangles, than one would expect if absolute space had a real physical effect. If it did, all angular momenta would be possible. In the case of systems subject to part II of the theorem, the system must have vanishing angular momentum.

Almost exactly forty years after Noether proved her theorem, Paul Dirac, who in the 1920s and 1930s had made some of the greatest discoveries in quantum mechanics, explored, in a different formalism and without mentioning any connection with the theorem, implications of its part II. Noether had noted that not all the equations of motion will be independent. Dirac picked up on the fact that at any instant not all the momenta in the equations derived from a variational principle of the kind Noether had considered in part II of her theorem are independent. This is perfectly consistent with Noether's observation since the equations of motion tell you how the momenta change in time. If the momenta are not all independent of each other, the equations that determine their change in time cannot be either.

Dirac called the conditions that at any instant restrict the independence of the momenta *constraints*. The conditions state that certain combinations of the momenta must be perpetually zero. In some of the most valuable parts of his work, he established the conditions under which theories in which this happens do not lead to contradictions—they are consistent. This means that if a certain set of constraints holds at an initial instant the equations of motion will ensure that they will hold at all times. The vanishing of the angular momentum in the relational three-body minimal model of Chapter 7 is a good example of such a constraint. It continues to hold under the evolution determined by the forces of universal gravitation because the Newtonian gravitational potential depends only on the separations between the particles and not on any orientation in space. In the development of the modern gauge theories of elementary particles, Dirac's consistency conditions play a critical role.

THE TAKE-HOME SUMMARY is that both Noether, in part II of her theorem, and Dirac explored—each with profound effect but in different terms—the consequences of one and the same state of affairs. It teaches us to be aware of immediate appearances. Like Newton, we think position in some uniform space is real, but in reality only relative positions can be observed. But then how is this to be reconciled with part I of Noether's theorem, which showed that conserved quantities reflect a uniform background? Is there or is there not such a background? I hope that you will already have an idea what answer I will propose. There is a background, but it is not the bland uniformity of Newton's absolute space. It's the universe. And it is history that links the two parts of mathematical physics, each architectonic in its own field of application, that Noether so brilliantly identified in her theorem.

The theory of the relational universe (Chapter 8) is a particularly simple example of a theory that meets the conditions of part II of Noether's theorem. This is because in it the action found by best matching has a value quite independent of where in space it is supposed to be determined. The value found is background independent. The connection between this and Dirac's theory is that for the relational universe the energy and angular momentum are exactly equal to the unique, special number that Schrödinger said has royal blood: zero. Their vanishing provides examples of constraints of the kind Dirac studied.

In the solution that gives rise to the 'most beautiful system' of Chapter 12, the universe in the region of the Janus point resembles a swarm of bees. In that melee there won't be any well-ordered behaviour in small isolated subsystems. A god-like observer would have to look hard and survey the whole swarm to find lawful evolution in it. But the correspondingly assumed divine abilities would eventually reveal that the complete universe has energy and angular momentum that vanish exactly not only at the Janus point but at all points in the solution. As just noted, these are the constraints, matching the conditions of a Noether II theory, that Dirac found: certain combinations of the momenta must vanish. Moreover, once the solution is examined sufficiently far from the Janus point, it will be found that subsystems are forming and becoming ever more perfectly isolated. What is more, in each subsystem one finds conditions like but not identical to the royal conditions—in each, energy, momentum, and angular momentum are becoming conserved with ever better accuracy.

If, as Schrödinger says, zero is the sovereign among numbers, the conserved quantities in subsystems are mere commoners. They are 'inferior' to the sovereign in two respects: first, they don't have royal blood (they are not zero) and, second, no matter the distance from the Janus point, perfect constancy is never achieved. But though the subsystems lack perfection, the universe would be dull without them. They are the stars and galaxies we see through our telescopes as well as the ideal gases in boxes studied by physicists. And all emerged out of chaos at the Janus point.

The mention of ideal gases brings us back to statistical mechanics. It was developed to a wonderful level to explain the thermodynamic properties of terrestrial systems in which certain quantities (above all energy) are conserved. Such systems exist under conditions for which part I of Noether's theorem holds. All attempts in that framework to give a genuine explanation of the second law of thermodynamics and the other arrows of time have failed. Has the problem been considered from a sufficiently general point of view? For all their brilliance, Clausius, Thomson, Maxwell, Boltzmann, and Gibbs had no chance of embedding their theory in a realistic model of the universe.

Because Noether II theories with a Janus point respect time-reversal symmetry, have of necessity architectonic arrows of time, and allow the emergence of systems with conserved quantities, even rudimentary models of that kind may be sufficient to model the universe, crack the riddle of time, and sketch a statistical mechanics of universes. The next step in this project, still for the case of nonvanishing size of the universe at the Janus point, is to justify the 'swarm of bees' state assumed there.

CHAPTER 14

A BLINDFOLDED CREATOR

THOSE OF YOU WHO HAVE READ STEPHEN HAWKING'S *A BRIEF HISTORY OF TIME* may recall that in the 1980s he started to devote much thought to a fundamental aspect of dynamical theories: they typically have infinitely many solutions. The question of which is relevant in any particular circumstance is seldom of much concern to experimentalists. They only want to predict what will happen if they set up an experiment with certain initial conditions. If the law is known, it will produce the answer. It's the same, for example, in observational astronomy. Given the positions and velocities of observed celestial bodies at a certain time, astronomers use Newton's laws to predict where the bodies will be many years later. The only difference when Einstein's equations are used is the greater complexity of the calculations. Such work is the bread and butter of much science and tends to condition the way its practitioners think. It does not involve and, as it were, puts on the back burner an existential question: if the law of the universe is known, which of its innumerable possible solutions are we in? That is the question.

Hawking was nothing if not ambitious and, in collaboration with James Hartle, put forth a daring proposal called the no-boundary condition. There's a moment in Wagner's *Parsifal* when Gurnemanz, who is leading the youthful hero to Montsalvat, the fortress where knights protect the Holy Grail, comments that they are making rapid progress because "here time becomes space". I have heard that Hawking was a Wagner fan. If so, he may have appreciated Gurnemanz's gnomic utterance; it captures rather well the essence of the no-boundary condition, which is that if we could go back in time to near the presumed start of the universe we would find that indeed time becomes space. Since a beginning of the universe presupposes time, this would mean the universe did not in fact have any singular beginning.

With the no-boundary condition, implemented in what has come to be called the Hartle-Hawking wave function of the universe and based on putative ideas about quantum gravity, the long-sought holy grail of theoretical physicists, Hawking hoped to kill two birds with one stone: first, create a theory of initial conditions and eliminate the inability of laws to determine actual solutions; second, as a bonus, show that the big bang would not be an unacceptable singularity at which critical physical quantities become infinite. It would not exist as the beginning of time since time would cease to exist and become well-behaved space.

I'm not going to describe the proposal of Hawking and Hartle, on which they worked more or less until Hawking's death. Certain mathematical issues, on which I think the jury is still out, made it controversial from the start. However, it does have the undoubted virtue of highlighting the issue of the connection between putative dynamical laws of the universe and the solutions they permit. I see three possible resolutions of the issue, all with the potential to advance our understanding of the universe.

The most radical is the one that, as I understand from a discussion with him a few years ago, the Nobel laureate Gerard 't Hooft favours. It is that the law of the universe and its solution are one and the same—a unique structure, as it were. My friend Lee Smolin, well known for his contribution to the development of loop quantum gravity and his popular science books, has also been attracted to this idea.

Next in the order of reducing possibilities is what I take the Hartle-Hawking proposal to be—not that there is only one possible solution but that one is singled out by a desirable feature shared by no other. That more than one solution is possible is implied by Hawking's name for his proposal: the no-boundary condition. The claim therefore is not that no other solutions exist but that only one has no boundary.

The next possibility is the one I propose in this chapter. The aim is not nearly as ambitious as the ones favoured by 't Hooft or Hawking. Rather, it is only to establish probabilities for solutions that have different properties. If it turns out that the overwhelming majority of solutions all share certain broad and distinctive features, this puts a new slant on the 'which solution' problem. You have not got much closer to predicting the probability of any particular solution, but at least you know what nearly all of the solutions will look like, perhaps in surprising detail. The law does more than might have been expected. Moreover, observations of the actual universe can indicate whether it is typical—that is, whether it belongs to the class of universes predicted to be highly probable.

Interestingly, Hawking together with his Cambridge collaborators Gary Gibbons and John Stewart also proposed exploration of this possibility in 1986 in a paper entitled 'A natural measure on the set of all possible universes'. In fact, the claim 'all possible' oversells the actual content; only a very restricted set of possible universes is considered. More problematically, even in its restricted domain the proposal had various defects. I will discuss the most serious one shortly. However, I mention the paper on account of its key idea. I find it most attractive; the form in which I present it will play a central role in this book from now on. If it does turn out to have a sound basis, that will be ironic, because it relies heavily on Boltzmann-type probability arguments, though in an inverted form—not to predict the equilibrated outcome of a thermodynamic process but to describe the initial state of the universe.

THE KEY IDEA is known by several names. The one used in the 1986 paper is Laplace's principle of indifference. It takes this form: if some process is known to have a certain number of possible outcomes and no further information is available, then if one is attempting to predict which outcome will be realised, equal probability of occurrence must be assigned to each of them. This, of course, is the only sensible strategy for gamblers betting on the fall of a fair die. As it happens, Laplace did advise gamblers; this may be what led him in one of his writings to state the principle, though he did not name it, perhaps because to him it seemed obvious. The name it now carries is due to the famous economist John Maynard Keynes. The caveat 'no further information' is clearly critical. If you know a die is not fair and that it is more likely to fall one way rather than another, even if only by a small amount, you would be stupid not to make your bet accordingly.

It should also be said that objections can be raised against the principle. By definition a fair die is one for which all sides have equal probability. However, the mere need to distinguish the sides of an actual die by distinct marks means that they are not all exactly identical. How might that impact their probabilities? Since the different solutions of a law are necessarily distinct, the same problem arises for them as for a die. Much more serious is the difference between a finite and infinite number of outcomes. In the latter case the very notion of probability becomes problematic. One can see this immediately by imagining that a fair die with infinitely many sides numbered 1, 2, . . . is about to be cast. I already suggested you would hardly want to bet on which number will come up.

When discussing the conditions under which the recurrence theorem holds, I noted that probabilities can be defined meaningfully in statistical mechanics despite the fact that in phase space there are infinitely many points, and hence infinitely many possibilities. This is because one can associate a definite volume, called the Liouville measure, with each well-defined region in phase space. In Chapter 10 I pointed out that, at least in principle, it can also be done in shape space because that space is compact. One can therefore ascribe to it a (nominal) total volume and, to regions within shape space, fractions of that volume. You will note that the word 'measure' appears in the title of the paper by Hawking and his colleagues. You will also note that the measure is claimed to be 'natural'. This is an implicit recognition of the problem that arises with probabilities for outcomes whenever there are infinitely many of them. Measures are not unique, so how can one be sure which measure should be chosen?

In fact, a first problem is to establish that the conditions under which a measure can be defined do exist. One could call this the problem of 'taming infinity'. Is there a natural way to do it? This is so in our case because the elimination of scale makes shape space compact. I hope you recall the parallel with Euclidean space (page 152): with its infinite extent, it is not compact, but the surface of a sphere embedded within it is. Shape space is analogous to such a sphere. This means that, in contrast to configuration spaces, in which overall scale plays an integral part, there are well-defined ways in which one can ascribe to shape space a finite size. But there still remains a second problem that I did not mention in Chapter 10 because it fits better here. It is an issue of physics: among all possible finite measures, which one has nature chosen? I argue in the notes that there are principles which do suggest a reasonably good candidate. I have called it the *neutral measure* and will discuss the issue of how we might check whether nature has chosen it later.

In the 1986 paper by Hawking, Gibbons, and Stewart, in which the elimination of scale plays no role, the problem is at the first, more serious level of taming infinity. The authors lacked a principle like the one that leads to shape space. Joshua Schiffrin and Robert Wald pointed this out in 2012. They used the graphic metaphor of a blindfolded creator throwing darts at an infinite board. Each point at which a dart can land corresponds to the creation of a possible universe and the answer to what I called the existential question: in which solution of the law of the universe, or at least in which kind of solution, are we likely to find ourselves? The metaphor highlights two problems: the blindfold and, more seriously, the dartboard's infinite extent. The second makes well-founded probabilistic predictions problematic, to say the least.

THE SETTING OF the problem is radically changed in shape space under the assumption that every solution has a Janus point. That makes the issue of initial conditions for the universe quite different compared with experiments in a laboratory or applications in astronomy. The last two relate to systems that are located in space and time relative to the background that the universe as a whole supplies. There is great freedom in where, when, and how initial conditions are specified. In contrast, for the kinds of universe we are considering, we will surely want to specify solution-determining data at the Janus point. This is the point that corresponds to the effective birth of time for each half-solution and is the unique point of balance at which the attractors identified in Chapter 11 do not yet impose their pervasive temporal asymmetry on each half.

Thus, at least in classical physics (before we attempt to take into account quantum mechanics), the 'where' and 'when' uncertainty about the specification of initial conditions is eliminated. Whatever nature chooses to do, that choice will be set in stone at the Janus point. That's built into the law of the universe. The question now arises of what kind of Janus-point conditions is likely to be chosen. Let's suppose nature follows Laplace's guidance for gamblers: all possible solutions are to be given equal probability. Unfortunately, this immediately leads to a difficult problem—that of determining the complete set of solutions and actual probabilities for them. I don't think the problem is insoluble, but let's see where the difficulty lies and where a partial solution might be found. We will find that surprisingly strong predictions may be possible.

The difficulty with determining the complete set of solutions arises from their inherent nature. Were the law of the universe geodesic, a point and direction in shape space would determine a solution. The rub, not yet seen to be a gift, is the mystery of scale (Chapter 8)—a geodesic law cannot give the magic of persistent structure growth. For that you need not only the point and direction but also the one extra datum from which all beauty and interest comes: the creation measure. It can have any value from zero to infinity—and so we are back to Schiffrin and Wald's dartboard.

For the moment let's ignore this problem and assume that Laplace's principle does at least apply to the position of the Janus point and the direction at it in shape space. If, in the absence of any better choice, we do assume nature uses the neutral measure, it is then a tractable problem to calculate probabilities for the complexity at the Janus point. Interesting predictions can also be made about the back-to-back directions there of the solution. Both of these possibilities are illustrated in Fig. 15. In our minimal model every solution must have a Janus point and back-to-back directions at it somewhere on the shape sphere. When darts are thrown by a blindfolded creator, the position at which each dart lands determines the Janus-point position of a possible solution, while the dart's fin determines the back-to-back directions from that Janus point. If the probability of a dart landing in a given region on the shape sphere is proportional to the region's area (calculated on the basis of the neutral measure used here), it is immediately seen that the dart is likely to land in a region of low complexity. The same kind of argument suggests that all fin alignments—that is, all back-to-back direction pairs of possible solutions at the pierced Janus point—are equally possible.

A mere glance at Fig. 15 shows that most of the shape sphere corresponds to shapes of low complexity. This effect, already manifested relatively strongly with just three particles, becomes significantly more pronounced with an increasing number of particles. The mathematical reason is given in the notes. Moreover, more or less spherically symmetric distributions are to be expected on statistical grounds. In anticipation of later discussion in Chapters

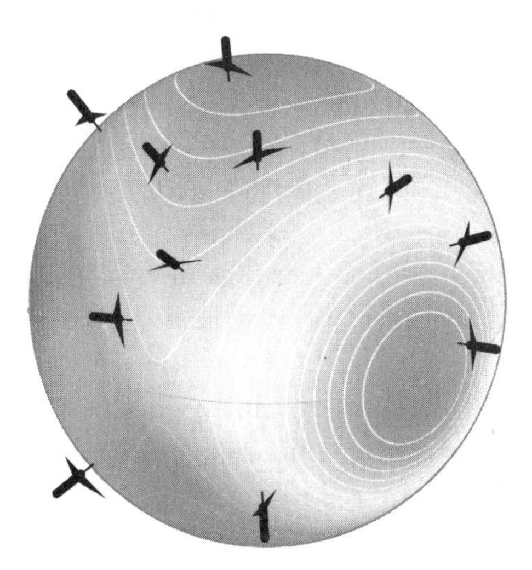

FIGURE 15. Darts thrown by a blindfolded creator at shape space.

16 and 18, it may be noted here that the points on the left of Fig. 11 correspond, as explained in the notes, to a Poisson distribution, while those on the right correspond to what can be called a glassy distribution. The former occupy the bulk of shape space, the latter a tiny fraction.

Of course, it is possible to question the blindfolded-creator argument, but it is striking that Laplace's principle of indifference, in the form adopted here, suggests that for any moderately large number of particles a high proportion of all solutions will have a rather uniform shape at their Janus points. They will be universes that seem to have been born in such a state. We will see a very significant enhancement of this effect in solutions in which the size of the universe vanishes at the Janus point.

THE LIKELIHOOD THAT the Janus point will be at a shape which is rather uniform and nearly spherical is the justification for the 'swarm of bees' initial condition proposed in Chapter 12. As regards their motions, I suggested that the bees would be flying in all possible directions with random speeds. I now want to suggest that the distribution of the speeds in the swarm could actually approximate, at least to some degree, a Maxwellian thermal distribution. This again is based on Laplace's principle of indifference, but now applied to directions in shape space. Here again we encounter the problem of a measure if we are to define probabilities meaningfully. Luckily the neutral measure we proposed for positions in shape space serves equally well for directions in shape space. On that basis we may ascribe equal probability to all directions.

A single direction in the shape space of *N* particles corresponds to the directions with which all *N* particles are moving in space. The one shape-space direction fixes the directions of their momenta but not their overall magnitudes, all of which can be increased or decreased by a change of units. If I made the point well enough, you will recall that all Maxwellian velocity (or equivalently momentum or kinetic-energy) distributions have a common dimensionless core and only acquire the different shapes shown in Fig. 1 (page 46) when units are chosen. To say it once more, the fundamental significance of the Maxwellian distribution is in terms of the relative proportions of the total kinetic energy distributed among the particles. In his example of just seven particles with discretised energies, Boltzmann showed that in the most probable macrostate three particles have no energy, two each have one-seventh of the energy, one has two-sevenths, and one has three-sevenths. This result is quite independent of the total kinetic energy available to the system. Of a single system that has nothing with which it can be compared, all one can say is that it has a thermal distribution; to say it has a temperature is meaningless. In contrast, two neighbouring systems can have well-defined different total kinetic energies; if their distributions are thermal, their temperatures will differ accordingly.

To repeat with some rephrasing the sentence that began the previous paragraph, a direction in shape space is shorthand for the distribution of the relative magnitudes of all the individual particle momenta. Boltzmann's great insight was that for a sufficiently large number of particles, overwhelmingly many of all the possible distributions correspond to thermal equilibrium. If, using the neutral measure, we apply this argument to the directions in shape space of the solutions that have their Janus point at a given shape, we can conclude, just as Boltzmann did, that the great majority of them will correspond to thermal equilibrium. The bulk of the 'universes' will seem to have been born with a more or less uniform spherical spatial distribution in space and a distribution of the particle energies that is approximately Maxwellian. The evolution away from the Janus point will immediately become subject to the attractors, whose existence is inevitable under the simple assumptions made about the nature of the model. Any observers within such a universe will see clusters begin to form and separate themselves from each other. The mean kinetic energy will be higher in some than in others, which will make it meaningful to begin to speak about a range of different temperatures in the various clusters.

In summary, on the basis of very few postulates, the two most important of which are expansion of the universe and a Janus point to ensure cosmic time-reversal symmetry, we have arrived at a story of the universe that inverts the one widely believed. It is not a story of fantastically finely tuned order at the birth of time that is then remorselessly degraded by the growth of entropy. Quite the contrary, it is the story of structure created out of chaos. And the mathematics that underlies it is mostly Boltzmann's, reformulated for a universe that is not in a box. What is more, in its broad features it resembles the universe in which we find ourselves.

Yet the story may take a twist. As I intimated when discussing Boltzmann brains, Feynman and Penrose could be right. The initial state of the universe might be extraordinarily special, though not through the addition of something to the known laws of nature. The law of the universe may enforce a Janus point at which the size of the universe vanishes. That brings with it a royal zero of immense power. There are two reasons to hold this twist back for one more chapter: first, it is more speculative than merely having aristocratic Janus points without a collapse to zero size, and second, finite-size Janus points already suffice to establish some secure results, above all the role-reversal effect due to expansion of the universe and identification of the origin of time's arrows. The royal zero sharpens their tips.

CHAPTER 15

AN ENTROPY-LIKE CONCEPT FOR THE UNIVERSE

Problems with the Conventional Notion

In the decades since the paradigm of the expanding universe became firmly established in the 1930s, there has been intermittent discussion of entropy in the cosmological context. As in the nineteenth century, it has generally been accepted that entropy can be defined for the universe and that, in agreement with the second law of thermodynamics, it should increase. In this chapter I will argue that the relational model suggests how an entropy-like quantity for the universe should be defined and that, contrary to the widespread expectation, it should *decrease*. There is no reason to dismiss this suggestion out of hand. The second law was established for systems that are confined, but much evidence suggests that the universe is not. Indeed, cosmologists are now convinced that its expansion is accelerating. Should the universe be described in the same terms as steam in a cylinder?

Conceptual confusion may be partly responsible for a too-ready belief that an entropy notion for the universe is unproblematic. The very idea that the universe encompasses everything means, almost by definition, that it cannot be affected by external forces. This is the condition which defines dynamically closed systems.

The statement that the entropy of a system either increases or stays constant is almost invariably made subject to this condition. Since the universe is assumed to be dynamically closed, increase (or constancy) of its entropy seems to follow.

However, the studies that led to the entropy concept presupposed a confining box with walls off which atoms and molecules were assumed to bounce elastically. Increase or constancy of the confined system's entropy presupposed the absence of any effect of external forces and no heat flow into or out of the box. Then the system, consisting of the atoms *and the box*, is dynamically isolated and thermally insulated. But the atoms are most definitely not dynamically isolated—they are subject to elastic repulsion whenever they approach the box walls. Without them there is neither equilibration nor entropy maximisation. It's as simple as that. Clausius unknowingly put walls around the universe when he said "Die Entropie der Welt strebt einem Maximum zu".

The main aim of this chapter is to define, for an unconfined expanding universe, an appropriate entropy-like concept. The adjective 'appropriate' is important. There is, it seems to me, a rather obvious way to define an entropy of the universe that looks like conventional entropy and will in general increase. However, the increase has a trivial origin and is not illuminating. Indeed, I will argue that the very nature of the increase casts doubt on the bona fides of the entropy candidate. The defect is easy to see. Suppose a box contains a dilute ideal gas whose particles interact through short-range elastic forces and have thermalised. Let the walls of the box be abruptly removed. The gas being dilute, there will be very few interactions between the particles before they fly off in all directions into the void. The distribution of their energies will remain close to Maxwellian, so there will be little change in the momentum part of their entropy. However, because the volume occupied by the collection of particles increases steadily, so will the configurational entropy. That looks like a steady increase of the entropy. However, the nature of the motion is transformed. The positions the particles have at any instant are determined by their initial positions and velocities in the box at the time when its walls are removed. Since the velocities almost instantly become constant and remain so, the positions of the particles are, after sufficient time, almost entirely determined by the velocities. Looked at from a larger perspective, the particles will seem to have emerged from an explosive event in a relatively small region, from which they are all streaming away radially and rectilinearly. What is more, the separations between any two particles are in accordance with the same law of increase of the intergalactic separations that Hubble found.

Thus, although the removal of the box walls does lead to a growth of conventionally defined entropy, the change in the nature of the motion does not at all look like the increase in disorder that is habitually associated with entropy increase. Quite the contrary, there is a transition from the epitome of chaos—the thermal equilibrium in the box—to what is just about the most ordered motion one can imagine. The inversion should surely raise doubts. In fact, I didn't really need the box to make this point. As explained in the notes to Chapter 6, particles that move purely inertially also have a Janus point. At it, the motions of sufficiently many particles will not differ much from the ones just supposed released from a box. That is also true of the Janus-point state assumed for the particles that evolve into the 'most beautiful system'. I rest my case.

In this chapter, I will propose an alternative notion of entropy which matches the intuition that a large value of entropy corresponds to a large amount of disorder. The big difference from conventional entropy is that, as suits a quantity which characterises the universe as opposed to subsystems of it, the entropy-like quantity is a pure number—as is the complexity, which plays the decisive role in the definition of the new quantity. The notion of entropy needs to be reconsidered if we are to apply it to the universe.

The Alternative

The main difference between the notion I am going to propose as entropy of the universe and Clausius's notion is that he obtained his by adding up the entropies of systems like those observed around us. They are all subsystems of the universe. Since most of them are effectively confined and the entropy in them does increase, the sum of their entropies will increase. The sum is what Clausius called the entropy of the universe. But he never addressed the fundamental question that bothered Boltzmann: "How does it happen that at present the bodies surrounding us are in a very improbable state?" It is because all these bodies, subsystems of the universe, are in improbable states that they have a low entropy and a universal tendency for that entropy to increase.

The quantity that I am going to propose answers Boltzmann's question. It is defined not as a sum of entropies of subsystems of the universe but directly for the universe as a whole, and not in the same way as conventional entropy. Its decrease reflects the fact that, as the universe expands, subsystems are of necessity formed in states whose conventional entropy is low and then increases. Familiar thermodynamic effects are observed, indeed explained, because it is expansion of the universe that causes the proposed entropy-like quantity to decrease. Thus, two complementary quantities must be considered: one for the universe, the other for the subsystems it creates. Thomson maintained you do not understand something in physics unless you can measure it and 'put a number on it'. The aim of this chapter is to put a number on the behaviour described in Chapter 12.

A comment on terminology: My collaborators and I initially called our concept the 'entropy of the universe', but the claim that it decreases created such opposition ("By definition, entropy must increase") that we switched to describing it as an entropy-like concept and called it *entaxy*. I will explain its etymology later; more important is that it is entropy-like.

Boltzmann's insight was that the number of a considered system's microstates which belong to a given macrostate defines its entropy. A macrostate, in its turn, is defined by means of one or more state functions. The volume and pressure of a system when in equilibrium are frequently used in thermodynamics, while in statistical physics the total energy is the most fundamental state function. All the possible microstates with a given value belong to a corresponding macrostate. By entropy-like I mean, as in Boltzmann's definition, that the chosen quantity is a count of the microstates that belong to a well-defined macrostate.

This might seem to be the standard approach and therefore should not be expected to lead to anything new. However, an important point, which I have not seen emphasised in the literature on time's arrows—in fact, I have not even seen it mentioned—is that, at least in classical physics, definite units of mass, length, and time are employed in all definitions of entropy used by physicists, chemists, and astrophysicists. It is true that in quantum physics entropy is expressed as the number of possible quantum states a system could have, but these too are defined by units, above all those of energy. There is freedom as to which units and means of their physical realisation are employed, but some must be. Critically, these are defined and exist independently of the investigated systems. A technician involved in a tabletop experiment of some process times it with a laboratory clock and picks up a ruler to make a measurement. Neither clock nor ruler is part of the process. This separation of system and experimentalist is deeply embedded in standard physics. In contrast, the universe 'does its own thing' with little apparent concern for physicists and their practice. I suggest we need a quantity that characterises the universe as it is, not as we measure it. In the *N*-body model, the complexity, the simplest pure number that reflects the relative distribution of its particles, is the most obvious quantity to choose.

Let's see how it compares with conventional state functions. Energy and angular momentum are the two state functions used in statistical-mechanical studies of stars and galaxies. Can they be used as state functions for the universe? For this purpose, they appear at first glance to have a desirable property, which is that they vanish. They have the value zero, and zero is zero in all units. We don't need units to quantify zero. It is a scale-invariant quantity.

The problem is that they both vanish at all stages of the evolution of our model universe. Although energy and angular momentum will be extremely useful once more or less stable subsystems have come into existence, as far as the universe as a whole is concerned they can have no value as conventional state functions at any stage of its history. This is because what is truly useful in statistical mechanics is to know how the entropy of a system changes as the value of a state function like energy is changed by energy input. This enables one, for example, to understand how and why ice melts when heated. But in a relational universe, the energy remains zero forever. Nevertheless, things manifestly change; ice melts without experimentalists pumping energy into the universe.

If we want to understand why things like that happen, we will need state functions that change. Moreover, their value must not depend on any extrinsic units of the purely nominal kind that Newton's absolute space and time made possible. Physically meaningful state functions for the universe as a whole must, like the complexity, be scale-invariant—pure numbers. In ordinary statistical physics, energy defines a microstate that simultaneously takes into account the positions and kinetic energies of the particles. I will first show how an entropy-like quantity can be defined for ratios of relative positions (shapes) in the relational universe.

Important here is the fact that, as realised in shape space, the universe 'makes its own box' (both in the case of *N* bodies and in spatially closed universes in Einstein's theory). It's the direct consequence of the removal of scale. With scale, the phase space of the universe has an unbounded measure; without scale, the measure is bounded. In essence, it's the simple difference between an infinite plane and the perimeter of a circle within it. This does not mean there will be recurrence in shape space; the dynamics in that arena is not characterised by standard frictionless Newtonian dynamics but includes effectively a friction term. The bonus of the 'shape-space box' is its rigour, both mathematical and physical. There are none of the worries associated with the idealisations that have to be assumed when treating subsystems of the universe. Without exception they are all imperfect. The universe, under the single condition of closure that so appealed to Einstein, is the one and only perfect closed system.

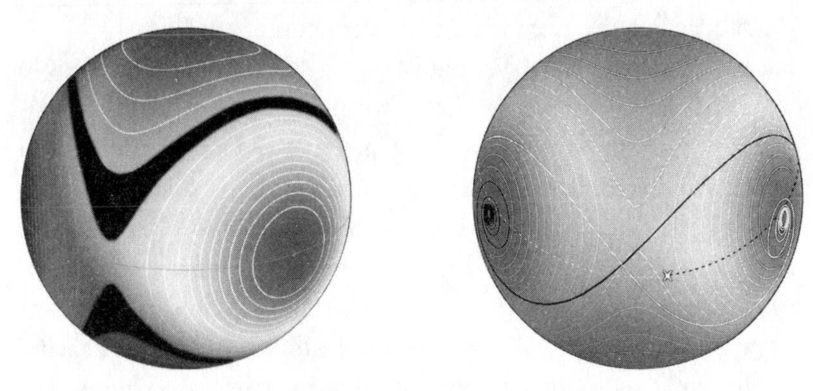

FIGURE 16. Definition of shape entaxy and its decrease.

As illustrated in Fig. 16, the notion of an entropy of the universe is modified accordingly. The shape complexity is chosen as the state function. Its equal-value contours are shown on both shape spheres in the figure. Macrostates can be defined in one of two ways. In the first all shapes with the same complexity belong to a corresponding macrostate. This could, for example, be the first contour that surrounds the 'north pole' in the figure. An alternative is to make shapes whose complexity values lie within a definite range—for example, the black region on the left—belong to a macrostate. Within macrostates defined in either of these ways, each point (that is, each shape) is a shape microstate. The entropy-like quantity, or *shape entaxy*, for the first kind of macrostate is the logarithm of the length of the equal-value contour that defines it; for the second kind it is the logarithm of the area between the defining contours. Individual microstates have the entaxy of the macrostate to which they belong. A Boltzmann-type conjecture is that the probability for a randomly chosen shape to lie within the black region is equal to the ratio of its area to the area of the complete shape sphere. The generalisation of these ideas to volumes of *N*-body shape space is straightforward.

However, the geometrical transparency of Fig. 16 hides a critical issue. It concerns the definition of distance (and with it area) on the shape sphere and the definition of volume in shape space.

As noted in connection with the representation in Fig. 13 of the shape sphere, there is a distinguished way of defining distances in shape space. I believe the grounds for using it are strong; the representation of the shape sphere in Fig. 16, as in Fig. 13, is based on it. The validity of any associated probabilistic predictions stands or falls with its choice. Therefore, if observers within a particular universe seek a probabilistic explanation of why they find themselves in that universe rather than any other governed by the same law, the choice is critical. This is just a restatement of the 'measure problem' discussed in Chapter 14. What is certainly true is that what observers within any possible universe can actually observe is unaffected by any theoretical choice of the measure in shape space—they will always experience an arrow of time. In the first place that reflects growth of complexity. That's the key concept. If there is a statistical notion of entropy for the universe, I feel sure it must be based on a quantity like complexity. There is a potential candidate for an analogous quantity in general relativity; we will come to it in a later chapter.

Every solution of our model universe has a Janus point. For the representative solution shown on the right in Fig. 16, it is at the small white cross at the 'back' of the shape sphere. As argued earlier in the book, this point represents the effective beginning of time for observers within the universe. They must be situated on one or the other side of the Janus point. In either direction from it the solution curve crosses complexity contours of higher and lower values but with a clear tendency for increase. There is a matching decrease in the lengths of the equal-value contours it crosses. Since the shape entaxies are proportional to the corresponding lengths of the contours, we see that the entropy-like quantity has a definite tendency to decrease in the direction of time defined by the growth of structure.

This is an appropriate point at which to explain the etymology of the term we chose, 'entaxy'. It has an obvious and deliberate similarity to Clausius's coining of 'entropy'. He justified '-tropy' by its meaning in Greek: 'transformation'. He added the prefix 'en-' to make the words 'energy' and 'entropy', which represent what he had identified as the two most fundamental concepts in physics, sound and look similar. Flavio Mercati's proposal of 'entaxy' uses 'en-' in the same way but also relies on its meaning of 'towards'. Combined with *taxis*, the Greek for 'order', this gives 'entaxy' the meaning 'towards order'. A state with high entaxy, acted upon by the law of the universe, has greater potential for transformation of chaos into structured order in the form of complexity than does a state with lower entaxy.

SO FAR WE have only discussed shape entaxy. In fact, the definition of the shape-space analogue of entropy associated with kinetic energy has already been anticipated in Chapter 14. It is associated with the direction of the curve in shape space. I argued that at the Janus point all directions can be assumed to be equally likely and that the majority of these directions will correspond to what looks like thermal equilibrium even though it will not be possible to associate a temperature with the state. Thus, something like Boltzmann's explanation of the Maxwellian thermal distribution should hold near the Janus point and permit a 'shape-direction' entaxy that has a near-maximal value.

However, with increasing distance along the solution curve we know that the attractor effect not only forces the system into special places in shape space (the wells of the shape potential) but also greatly limits the possible directions in which it advances. The number of direction microstates must decrease. This is already manifest in the way the trajectory spirals into the wells of the shape potential in Fig. 16. Among all possible directions of the curve, only a tiny fraction are realised. For example, there are none anywhere near the radial directions away from the bottom of the well. Thus not only the shape entaxy but also the direction entaxy will decrease. It's a direct consequence of the attractor effect's iron grip.

The conclusion of this chapter is therefore that any sensibly defined entropy-like quantity for the model universes we are considering will decrease away from those universes' apparent beginning in time. This reflects the way in which such universes become increasingly structured. Moreover, the inclusion in more realistic models of other forces, above all electromagnetic and nuclear, as well as the effects of quantum mechanics, supports this general picture of structure formation in our actual universe. Indeed, observations of it from the earliest epochs to the present time reveal a picture of matter clumping as the universe evolves, this happening on a huge range of scales—from protons and neutrons right up to clusters of galaxies. This encourages me to think that the basic picture of decrease of an appropriately defined entropy-like quantity for the universe is correct at least up to now in the history of the universe and possibly far into the future if not forever. The reconciliation of this conclusion with the second law of thermodynamics is the subject of Chapter 19, but it is worth presenting some of the background to it now, above all the part played by black holes.

Black Holes and Their Entropy

Many years passed between Einstein's creation of the general theory of relativity in 1915 and the clarification of at least some of the most important properties of black holes. Early difficulties were associated with separating artifacts of mathematical description from physical facts. Only in the 1960s did real clarity begin to emerge, partly stimulated by increasing evidence for the existence of concentrations of great amounts of matter in very small regions within galaxies. Many theoreticians, in particular Roger Penrose, contributed to the work of clarification. Pedro Ferreira's *The Perfect Theory* gives a particularly good description of this work; also highly recommended is *Black Holes and Time Warps* by Kip Thorne, who has recently been awarded the Nobel Prize in Physics for his outstanding contribution to the discovery of gravitational waves and as a student was one of the richly talented group at Princeton around Richard Feynman's 1940s mentor John Wheeler.

The most important well-established fact about a black hole is that it has an event horizon. This is two-dimensional and is the fateful surface through which you can pass but through which not even light can escape (at least before quantum effects are taken into account). The event horizon is traced out in spacetime by light rays that just escape being captured by the black hole; they skim, as it were, along the event horizon's edge. One of the greatest successes of observational astronomy, achieved in 2019 by coordinated observations by radio telescopes at various locations on the earth, was the image of the event horizon of the giant galaxy M87's black hole of 6 billion solar masses. The event horizon is approximately 40 billion kilometres across. If the sun were to become a black hole, which it cannot because it is not sufficiently massive, its event horizon would be about 2.5 kilometres across (the sun's current diameter is about 1.4 million kilometres). One of the puzzling things about event horizons, especially of very massive black holes, is that if you were to pass through one of them, nothing untoward would seem to have happened. You would, however, be bound inexorably on a course to the singularity at the black hole's centre, where brutal tidal forces would pull your body apart.

As the properties of black holes became better understood, people began to wonder about their connection with entropy and the second law of thermodynamics. Suppose one were to drop a box full of radiation, with well-defined entropy, into a black hole. The entropy of the universe outside the hole would decrease, in apparent violation of the second law. It was also noted that black holes had certain similarities to thermodynamic systems. In equilibrium the latter are described by just a few parameters—pressure and volume in the case of an ideal gas. A black hole for its part is fully described by its mass, its angular momentum, and whatever charge it might carry. Struck by this bare simplicity, John Wheeler, never short of the *mot juste* and already coiner of the now ubiquitously invoked 'black hole', capped it with the epigram that a black hole has no hair. It dawned on researchers that black holes seemed to share with 'common or garden' equilibrated systems the status of being the end state of a dynamical process.* More intriguing was Hawking's proof that any mass which fell into a black hole

---

*Indulging my interest in etymology, I discovered that the expression 'common or garden' may have arisen through reference to a plant often found on common land or in gardens.

increased the area of its event horizon. This looked like entropy increase.

These formal similarities with thermodynamic systems were an enigma: thermodynamic relations had long ago found their nutshell statistical-mechanical interpretation. How could the concomitant probability find its way into Einstein's entirely deterministic theory? There could not be anything more than a mere formal analogy. However, in 1973 John Wheeler's PhD student Jacob Bekenstein argued on intuitive grounds that the formal similarity had a physical foundation and that black holes must have a genuine entropy. Stephen Hawking was sceptical and, intrigued by a suggestion of the Russian physicist Yakov Zel'dovich, started to make calculations.

He first had to teach himself quantum field theory, the most advanced form of quantum theory and already difficult in the absence of gravitational effects but particularly tricky in general relativity's curved spacetime. In an exercise worthy of a trapeze artist—it took him from a quantum state that existed before a black hole's creation to one to which the event gives rise in the distant future—Hawking, bounded like Hamlet but in a wheelchair, begot a sensation, first among the community of relativists but then rippling around the world. Far from refuting Bekenstein, he proved him right; indeed, he showed that the analogy between thermodynamic systems and black holes went significantly further than hitherto suspected. For all his daring, Bekenstein had shied away from associating a genuine temperature with black holes and had only spoken of an effective one. Hawking, in contrast, found that black holes must have a real temperature proportional to their surface gravity. This is what you feel when standing on the earth—the gravitational pull toward its centre. For a black hole it's the thrust a spacecraft hovering just above the event horizon must maintain to avoid being dragged through it to inevitable destruction.

Alongside the temperature, completing the analogy with thermodynamic systems, Hawking showed that a black hole does indeed have an entropy, which can be expressed in two equivalent ways. First, besides depending on Newton's gravitational constant and the speed of light, it is proportional to the square of the black hole's mass and inversely proportional to Planck's constant. Since this last has a tiny value, the entropy of a solar-mass black hole is huge. In fact, the total entropy of the black holes known to exist within the observable universe dwarfs the entropy of all the matter outside them. Second, and as Bekenstein had conjectured, the entropy is proportional to the surface area of the event horizon. This is particularly remarkable since conventional entropy is proportional to the volume of the considered medium. Hawking was able to make precise the estimate of the black-hole entropy that Bekenstein, forced by the absence of any other suitable universal constant, had obtained by his introduction of Planck's constant, which has the same dimensions as angular momentum and, introduced into his calculations, made it possible to obtain the correct mass, time, and length dimensions for entropy. This was the second time in the history of physics that dimensional analysis, introduced by Joseph Fourier in 1822 to ensure that both sides of an equation have the same dimensions and developed into a fine art by Maxwell, played a decisive role; moreover, both times were in a quantum context (in Planck's original discovery in 1900 of the role his constant plays in the thermal equilibrium of radiation and then in Bekenstein's hazarding of a black hole's entropy).

Unifications of the apparently distinct are turning points in physics. The first occurred when Newton brought terrestrial and celestial motions into a shared dynamical framework. Next Maxwell showed that the speed of light tied together magnetism and electricity. Then, half a century or so later, Einstein dissolved the difference between gravity and inertia. Bekenstein and Hawking reached the current high-water mark in this process in unification by linking in an indissoluble unity the three most fundamental quantities in physics: the speed of light, Newton's gravitational constant, and Planck's constant of action. This latest advance gave strong support to Bekenstein's already proposed generalised second law of thermodynamics: the total entropy of the universe either remains constant or increases because any entropy lost by regions outside black holes—for example, when a box of radiation is thrown into one—is offset by an equal or greater increase in the black-hole entropy. Another consequence of the discovery that black holes have a temperature is that they must slowly evaporate through what is now called Hawking radiation. For a massive black hole, the temperature is extremely low and increases slowly as the hole evaporates. The evaporation time is immensely long, very much greater than the age of the universe for solar-mass black holes and much, much longer again for the extraordinarily massive black holes that exist at the centre of many, perhaps all galaxies. What happens in the final stages of the evaporation process is unknown; that will probably become clear only when a definitive theory of quantum gravity, the unification that some believe will complete physics, has been achieved.

Since that unification is the goal—and graveyard—of many careers in physics, it is not surprising that Hawking's result unleashed a flood of theories and conjectures; they continue unabated. Virtually all theoreticians are convinced that the Bekenstein-Hawking entropy is a key that will unlock the door to quantum gravity. But the way to turn the key has not yet been found. One idea that has gained widespread support is the holographic principle. Its basis is the surprise that black-hole entropy is proportional to an area, the surface of the event horizon, while conventional entropy is proportional to the volume of a system in equilibrium. Given the connection in ordinary statistical mechanics between the number of information-bearing microstates of a system and the volume that system occupies in equilibrium, it was natural to suppose that analogous information is encoded on the area of a black hole's event horizon. The holographic principle takes the conjecture much further; it proposes that all the information contained in a volume of space is encoded on the surface of that volume. Despite this exciting idea, for which partial realisations do exist, as yet none of the ideas advanced for the form of quantum gravity have broken through to anything widely recognised as fully fledged quantum gravity. Much work, including the great amount done in string theory, has a hybrid nature in which quantum excitations are described on a classical spacetime background. That's not a purely quantum description; it's not background-independent. Loop quantum gravity, the main rival, did attempt such a theory, but despite initial promise it has failed to resolve the great problem of the apparent complete disappearance of time in quantum gravity, the subject of my book *The End of Time*.

The discovery that black holes have entropy has a dual significance in the context of this chapter. First, it is consistent with entaxy decrease being an indicator of conventional increase of the universe's entropy calculated in the way Clausius did, as the sum of the entropies of its subsystems. The point is that the formation of a black hole corresponds to the concentration of a great amount of matter in a very small volume. This will greatly increase the complexity of the universe and hence decrease its entaxy. At the same time, the Clausius-type entropy increases—massively, in fact. However, it is this very increase which seems to provide strong support for the standard view that the universe has an entropy and that it increases. Through Bekenstein's generalisation of the second law it also extends to gravity the domain of phenomena subject to the rule of entropy increase.

The person who, since 1979, has done more than anyone else to develop influential ideas about gravitational entropy, in general situations and not just in black holes, is Roger Penrose. Two things were his stimulus: first, the Bekenstein-Hawking result; second, observations in cosmology which suggested rather strongly that the universe was in thermal equilibrium at the big bang or at least very soon after it. That was a real puzzle, because if the universe had already begun in a state of thermal equilibrium, for which the entropy of matter is maximal, that did not seem to leave any possibility for further entropy increase, as required by the second law.

Penrose found himself forced to adopt a proposal like the one Feynman had made about fifteen years earlier—namely, some additional condition of extremely low entropy not enforced by the hitherto known laws of nature must be imposed at or very soon after the big bang. The difference now was that, unlike Feynman, Penrose was prepared to entertain a conjecture about the nature of an extra condition of this kind. Like others before him, he took seriously various hints that gravity is 'anti-thermodynamic'. In particular, whereas non-gravitating matter confined in a box tends to a uniform equilibrium distribution, gravitating matter when confined will tend to clump; indeed, in accordance with the expression for black hole entropy, the maximum entropy for matter in a box is achieved when all the matter has collapsed into a single black hole.* In accordance with this line of argument, a uniform matter distribution, which in normal statistical mechanics corresponds to maximal entropy, would represent a state of minimal gravitational entropy. Then, even in a state in which the entropy of matter is maximal, the overall entropy of the universe could, in view of the potential of black holes to generate a vast amount of entropy, be very low and have ample opportunity to increase as black holes are formed.

Penrose made no attempt to propose a mechanism by which such a low initial gravitational entropy of the universe could come about, though he did conjecture that it could be due to the as yet unknown theory of quantum gravity. A further problem that has hitherto remained unresolved is what expression can be proposed for the entropy of the gravitational field of the universe between the initial state of very low entropy at the big bang and the formation of the first black holes. Despite these difficulties, a not insignificant number of theoreticians have found Penrose's proposal persuasive even though black hole entropy remains enigmatic, since no proper theory of quantum gravity yet exists. Consider a neutron star with mass just less than what is needed for it to collapse and become a black hole. Let it accrete an ounce or two of matter. From a state with modest entropy the star is mysteriously transformed, almost in a flash, into a black hole with mind-boggling entropy. How does that happen? Nobody knows for sure, but many believe it is due to the effect of quantum entanglement, about which I will say something in Chapter 16.

---

*Arguments along the lines just indicated that gravity is 'anti-thermodynamic' because of the associated tendency to clump strike me as questionable because they rely on the artificial assumption that the self-gravitating system is confined to a box. But self-gravitating matter, even when some of it collapses to form a black hole, is never in a box and some matter can always escape collapse.

CHAPTER 16

A GLIMPSE OF THE BIG BANG?

Total Collisions

This is the means of obtaining as much variety as possible but with the greatest order possible; that is to say, it is the means of obtaining as much perfection as possible.

—GOTTFRIED WILHELM LEIBNIZ

IN 1714, TWO YEARS BEFORE HE DIED, LEIBNIZ PUBLISHED HIS *MONADOLOGY*, from which the words at the start of this chapter are taken; they summarise his philosophy in two sentences. In *Candide*, Voltaire mocked the long-dead Leibniz, in the form of Dr Pangloss, for claiming we live in the best of all possible worlds; a more tenable thesis is that variety, expressed through complexity, increases without bound. This chapter considers evidence for it in a Newtonian theory of the big bang; Chapter 17 will consider whether a similar realisation is possible in general relativity. If nothing else, the Newtonian big bang is of great interest, not least because it explicitly provides, as Feynman and Penrose conjectured, a very special initial state for the universe. However, I do not think such a state is needed to ensure universal validity of the second law of thermodynamics;

expansion of the universe on either side of a Janus point will do that. The state I will describe does that too but has implications which go far beyond the second law and can, I think, be argued to be the most direct consequence of a maximally predictive law of the universe rather than a past hypothesis invoked ad hoc.

Intriguingly, the solutions that it allows exhibit some similarities to quantum mechanics that one would hardly expect in a purely classical model. This is all because the size of the universe at the Janus point is no longer finite but zero. The possibility that the size does go to zero has long attracted the interest of *N*-body theoreticians on account of its exceptional nature. It appears that Newton's equations break down at the point of zero size because no way has been found to continue their solutions past that point. At it, there is a *total collision*: all the particles collide simultaneously at their common centre of mass. Its significance for us is the remarkable manner in which the system approaches the singularity, as well as the special shape, called a *central configuration*, that it has there; it is still well defined despite the vanishing size.

At this point I am going to clarify terminology to match the different temporal interpretations one can put on the same timeless reality of curves in shape space. If all the particles are at their common centre of mass, I will say there is a *total coincidence*. When that possibility is realised, I will, when appropriate, say there is either a *total explosion* out of total coincidence or a total collision at it. The alternative of total explosion anticipates the use of such a situation to model the big bang, which is permissible since time-reversal symmetry means time can be supposed to go in either direction. If we define the direction of time by growth of complexity, as we have hitherto done, there is no doubt: experienced time increases away from the point of total coincidence. The historical context, strongly influenced by the Newtonian concept of absolute time that flows unstoppably from past to future, led *N*-body theorists to consider the circumstances under which initial conditions formulated in the traditional manner might lead in the future to a singularity in the form of a total collision. I will retain that terminology when it reflects more faithfully the historical discovery but use 'total explosion' when thinking of the big bang.

With this settled, the first thing is to give you the necessary details about total collisions. They occur in the very special solutions mentioned in Chapter 7 but not described there when I introduced the minimal model. They can happen in two ways. In one the shape of the system, which must be one of the special central configurations, remains unchanged; such a motion is said to be *homothetic*, from the Greek *homothésis*, which is best translated as 'same arrangement' because the shape which the particles form does not change during the motion. Such motions are relatively simple, but the proof that they exist is not. In the three-body problem, Euler discovered the first examples. In them the three particles are always on a line and the configuration they form has throughout the shape of the Euler configurations on the equator of Fig. 13. Each of the three particles can be between the other two in a position determined by the mass ratios. Lagrange found a fourth and most striking example and published it in his 1772 paper. In it, the three particles are, for any values of their mass ratios, at the vertices of an equilateral triangle. Those four are the only examples associated with homothetic total collisions in the three-body problem.

The really interesting total collisions—the subject of this chapter—are nowhere homothetic but reach a central configuration in a very special way that allows a total collision to occur. In a remarkable paper in 1913, Karl Sundman, a Finnish mathematician of Swedish origin, proved that such solutions exist in the three-body problem; in 1918, in an even more remarkable paper, the Frenchman Jean Chazy proved they exist for any number of particles in the N-body problem. They can occur only if the angular momentum is exactly zero, but there is no restriction on the energy. With time run in the direction away from the point of zero size, such solutions are, as I just said, total explosions and models of a big bang. Sundman's paper appeared two years before Einstein created general relativity, and Chazy's was published eleven years before Hubble discovered the expansion of the universe! However, I am not aware of much comment on this Newtonian precursor of the Einsteinian big bang. Here I will argue that the '*N*-body big bang' warrants discussion, above all because it hints at hidden depths in general relativity, in which a large class of solutions have zero-size big bangs; there is also the possible connection with quantum mechanics I already mentioned. For these two reasons the Newtonian big bang suggests new avenues of research and is simultaneously an example—I think it is the first that has been proposed—of an explicit past hypothesis. However, its form suggests, to me at least, that it is not something that needs to be added to the laws of nature but rather is a direct consequence of a hitherto unrecognised law of the universe from which the laws themselves are emergent.

As I have already said, total collisions can occur in shape space only at special points called central configurations. In connection with the importance I have attributed to complexity, it is particularly interesting that central configurations exist only where complexity has an extremum. This is an arrangement of the particles such that any infinitesimal change in that arrangement does not lead to any change in its complexity. There are innumerable examples in mathematics of extrema: maxima, minima, and saddles. The last of these, as the name suggests, are like the saddles put on horses: flat in the middle, sloping down on two sides to the stirrups and upwards towards both the horse's head and tail. For a maximum, imagine cutting a perfectly spherical earth through the equator and placing the Northern Hemisphere on a table with the pole at the top. From the pole the height of the surface decreases smoothly at the same rate in all directions. That's a maximum. There will still be a maximum as long as the height decreases in all directions. Now turn the hemisphere upside down and balance it on the pole; the height of the surface will increase from the pole in all directions. The pole is now at a minimum. Extrema of the same three kinds exist in spaces of any number of dimensions greater than two; in them saddles can have different numbers of directions of ascent and descent. The complexity has no maxima, only sharply pointed infinitely high peaks; it is the minima that are really interesting, especially the one with the lowest complexity, which in general is unique. The shape potential is the negative of the complexity and has maxima corresponding to the complexity's minima. Bear with me if I continue to use both names, for one and the same mathematical formula characterises two quite different things: structure and the potential for change through gravity. Both are expressed in purest form at an extremum.

In the three-body problem the complexity has four extrema. There are saddles at each of the three collinear Euler configurations; the fourth extremum, the most interesting one, is a minimum in which the three particles, whatever their masses, form an equilateral triangle. They correspond to the shapes of the four possible homothetic motions just described. In the late nineteenth century it was discovered that a regular tetrahedron has the same property. In it, four particles with arbitrary ratios of their masses are all equally spaced from each other and form a pyramid, with three particles as an equilateral base and the fourth particle as the tip of the pyramid. These are rare and atypical examples of exactly symmetric central configurations. Any central configuration, being an extremum of the complexity, has the same value for its complexity no matter how it is increased or decreased in size and oriented and located in space; Fig. 10 (page 141) illustrates this for the complexity of an arbitrary configuration, and it must remain true for an extremum. Whereas in the three-body problem there are only four central configurations, the number increases very rapidly with the number of particles and may not even be finite but form a continuum of possible shapes.

It is important to note that it is the mass ratios of the particles, nothing else, that determine the possible shapes of central configurations. Fig. 17 shows three planar equal-mass eight-particle central configurations together with the values of their complexity in the top line and the differences in the bottom line showing by how much the corresponding shape potentials are lower than the maximum; the one with its lowest value is perfectly symmetric, while the other two (you will have to look quite hard to see the difference between them) have only near left-right symmetry. Especially for central configurations with complexity near its minimum possible

FIGURE 17. Three planar central configurations of eight equal-mass particles.

value, very near but not perfect symmetry is a typical property. The effect is illustrated dramatically in Fig. 18 for a 500-particle central configuration with minimal complexity. The reason for its near-perfect sphericity is explained in a paper mentioned in the notes for this chapter. I have a conjecture, to be presented in Chapters 17 and 18, that an analogous effect in general relativity is the origin of all structure in the universe. This chapter presents initial supporting arguments.

Central configurations are so named and lead to total collisions because in them the combined gravitational forces exerted on any of the particles by all the others are directed exactly toward the centre of mass of the complete system. As if by some conspiracy, the forces between the particles that otherwise always change the shape of the system disappear. If one imagines all the particles being held at rest in a central configuration and then released, they

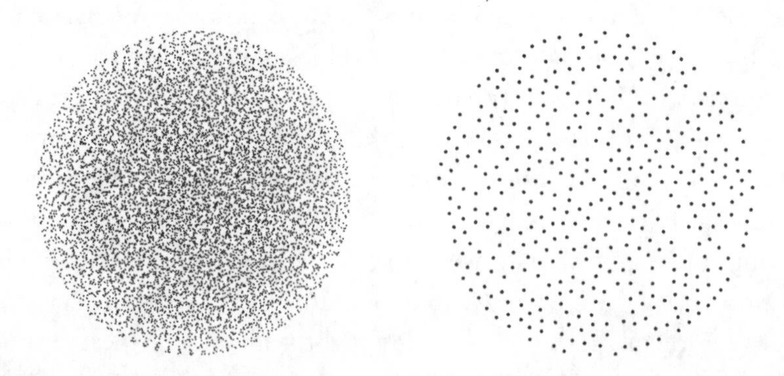

FIGURE 18. Near-maximal central configuration of 500 equal-mass points with slice through it.

will all start to fall in straight lines to their common centre of mass, where a total collision occurs. The dashed lines in Fig. 19 (please ignore the solid curves for the moment) show how this happens for particles with masses having the fractions 1/6, 1/3, and 1/2 of the total. They start their fall from rest at the vertices of an equilateral triangle, with the lightest particle furthest from the common centre of mass and the heaviest nearest it. The successive equilateral triangles show the size of the system at equal intervals of Newtonian time. The increasing speed of the motion is reflected in the increasing separations. The figure demonstrates several features of Newtonian gravitational dynamics rather well. First, the effect of the mass ratios, the 'gold-standard' pure numbers that characterise the system, is clear. The heaviest particle, with mass ratio 1/2, comes in from the bottom right. It is close to forming one side of a seesaw with the other two particles having the point of balance at the centre of mass. Like a parent balancing a child, it starts closer to the point of collision and therefore can move slower than the other two and still be on time to ensure that the three meet simultaneously at the centre of mass. To do so each must speed up its motion, initially gradually and then more rapidly as it gets closer to the point of collision. The somewhat delayed acceleration reflects the $1/r^2$ dependence of Newton's force law. Its effect is weak at first but becomes unstoppable as the particles approach the point of collision, where their speed becomes infinite.

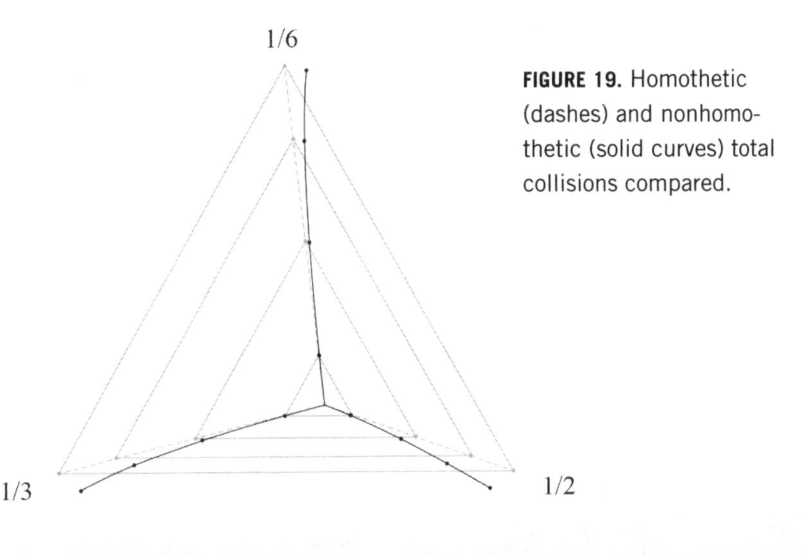

FIGURE 19. Homothetic (dashes) and nonhomothetic (solid curves) total collisions compared.

The example I have just described, with the particles at rest some distance from the centre of mass, corresponds to a system with negative energy. We can also suppose that it begins at zero size with a total explosion and reaches a maximum size where it comes to rest before falling back to a total collision. If the energy is exactly zero, the particles that emerge from a total explosion keep on moving apart forever but slower and slower because the potential energy decreases continually in magnitude; by energy conservation, the kinetic energy must decrease as well, and with it the particles' speeds. The three particles get to infinity with their last gasp, with zero speed. If the energy is greater than zero, each particle effectively escapes from the gravitational attraction of the others. The motions become almost perfectly inertial and all three particles 'coast to infinity'.

Before we consider nonhomothetic total collisions as depicted by the solid curves in Fig. 19, some further comments about the homothetic solution need to be made. Behaviour of the kind it illustrates requires perfect fine tuning and cannot be realised within the universe, which will always exert some disturbance. The depicted solution is almost entirely reliant on representation based on Newton's imagined absolute space, time, and scale. Without scale, one cannot give the triangles different sizes, for nothing but shape remains; without time, you cannot put changing amounts of duration between them; without space, they could be laid around anywhere. The three different scenarios corresponding to negative, zero, and positive energies become fictions. Vonnegut's Rockies and the spacings between them are real, like the triangle shapes of the minimal model I said my grandson could order using the shapes alone; because there is no change of shape in the homothetic solution in Fig. 19, the only reality Jonah could take from it is an equilateral triangle. The imagined Newtonian framework in which it is depicted is seriously deceptive.*

---

*But it must be said that extraordinarily beautiful things happen in it. At the time of writing a splendid example can be found in the Scholarpedia article 'Central configurations', by Richard Moeckel, who is one of the world's leading *N*-body specialists (www.scholarpedia.org/article/Central-configurations). Its Fig. 6 shows

In Chapter 17 we will see that this has an important bearing on the way cosmologists conceptualise, but it can be noted here that, as regards the manner in which the dynamical evolution unfolds, the three possible three-body homothetic total explosions (using now the more appropriate terminology) are exact one-to-one counterparts of the simplest (involving only dust particles) expanding-universe models that have now played an important role in cosmology for close on a century.

The problem that arises from this can be adumbrated here. Several chapters (Chapter 7, on the minimal model; Chapter 8, on Machian dynamics; and Chapter 12, on the 'most beautiful system') showed how, far from the Janus point, the law of the universe generates emergent subsystems together with the framework in which their dynamics can be described. Without the framework, all that remains of the homothetic solution in Fig. 19 is the bare notion of an equilateral triangle, a mere point in shape space. Nothing happens. The danger is that something whose existence is created within a context is used to model the context itself (the universe). The model is the nonhomothetic collision, which we now consider. The dots on the curves that the three particles trace out mark on them four precollision instants in the evolution. They exhibit genuine change of shape. Superimposed on that reality we see, suitably in grey as an 'imagined spectre', the homothetic collision, with one and the same equilateral triangle depicted with different sizes. Let us now banish the spectre and see what reality looks like. That will need some further description of the structure of shape space.

---

a homothetic three-body total collision but also a much more remarkable video of *homographic* motion (its Fig. 8) in which seven particles always have the shape of a central configuration (its Fig. 3) that 'breathes' in and out as it rotates (for such motions to exist, they must be planar—all particles in a plane). The constant non-vanishing angular momentum in the system prevents it from collapsing to a total collision, but each of the seven individual particles traces out, around the common centre of mass, an ellipse of the same eccentricity, 0.8, but different size. All particles move in exact phase with each other, and the whole spectacle is a perfect ballet because the motion mimics the ballerina's piroutte, the particle collective turning faster and slower as it 'breathes' in and out. Even William Blake, with his intense dislike of Newton, could hardly fail to wonder at what his laws can bring about.

The Topography of the Universe

Fig. 20 shows two shape spheres viewed from above the north pole; the one on the left is the equal-mass shape sphere you have already seen more than once, but now on the right is the one for particles with mass fractions 1/2, 1/3, and 1/6. Clockwise from the top right for this shape sphere, the dark regions without contour lines (too cramped to be shown) are the potential wells corresponding to close proximity of the (1/2, 1/3), (1/2, 1/6), and (1/3, 1/6) particle pairs. While there is still mirror symmetry of the two hemispheres in the unequal-mass case, the maxima of the shape potential (minima of the complexity) are no longer at the poles but, as you see, somewhat displaced. For each shape sphere the extremum of the shape potential, the point of possible total coincidence, is in the middle of the white central patch. With the equal-value contours that the unequal-mass complexity defines on it, the shape sphere acquires an asymmetric, somewhat lopsided appearance.

At the end of Chapter 9, anticipating the content of this chapter, I said a coincidence of particles that at first seems to be a frog may turn out to be a prince. In fact, the unequal-mass shape sphere looks to me more like a three-legged frog! The projection used is such that the radial distance from the centre of the image is proportional to the difference of latitude from the pole. Ridges of the shape potential run down to the Euler configurations from the maxima of the shape potential; between them are 'scree slopes' extending to the three possibilities for two of the particles to be coincident. Fifteen contours of the shape potential from its summit are shown; they should continue into the black regions but, as just noted, get too cramped to be shown. Because of the projection employed, the true spaces between the contours cannot be faithfully depicted. Nothing distinguishes the slopes and ridges in the equal-mass case on the left, but there is a clear ranking in the unequal-mass case. This has important consequences, to which we now turn.

When I introduced central configurations, I emphasised that it is mass ratios and nothing else that determines their position in

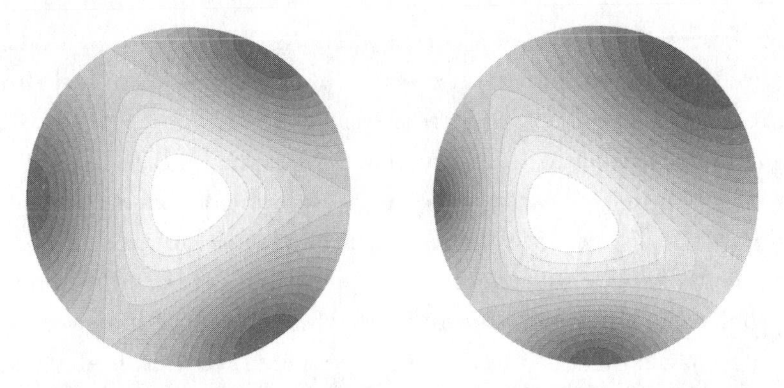

FIGURE 20. Two 3-body shape spheres viewed from above their north pole.

shape space. It also determines the 'topography' that the equal-value contours of the shape potential define in the whole of shape space and, critically, in the immediate vicinity of the extrema at which central configurations are located. On the two-dimensional shape sphere there are just two possibilities: either the initial rate of decrease away from the summit is exactly equal in all directions or there are two back-to-back directions of steepest descent together with another pair of least descent. This is what one has on a rugby ball. If the ball is balanced on one end, all directions from the other end have the same curvature and initial rate of descent of the ball's surface. However, if the ball is laid flat on the grass, then at the highest point of its 'waist' the curvature, and with it the initial rate of decrease in height, is least towards the two ends of the ball and greatest around the waist. The two possibilities corresponding to the end and waist of the ball are realised for the two shape spheres of Fig. 20. In the equal-mass case, the initial rate of decrease is the same in *all* directions; this is a consequence of the tripartite symmetry that results from the equality of the three masses. When the symmetry is broken because the particles have unequal masses, the second possibility, of back-to-back directions of steepest and least descent, appears in the topography of shape space.

Mathematically, the situation I have just described is encoded in what is called a Hessian, which characterises all kinds of extrema and exists at any central configuration not only in the two-dimensional shape space of the three-body problem but also in the $(3N-7)$-dimensional shape space of the N-body problem. If, as I shall assume, there is no symmetry, then the Hessian of the extremum in shape space that we are considering defines a matching number of distinct special directions, called *eigendirections*.* They are actually back-to-back directions like the two in the three-body problem. In the four-body problem, which has a five-dimensional shape space, there are accordingly five such directions; in the five-body problem, there are eight; and so on, increasing by three for each extra particle. They can be arranged in a sequence, starting with the one that has the maximal initial rate of descent down to the one with the smallest. The initial rates of descent are determined by what are called the *eigenvalues* of the Hessian. There is one of these for each eigendirection. Another important thing is that just as the possible shapes of the central configurations are completely determined by the mass ratios and nothing else, exactly the same is true of the eigendirections and eigenvalues of the Hessian. I should also say that Hessians with distinct eigendirections and eigenvalues are the norm. Even when all the masses are equal, very few central configurations are exactly symmetric.

In the unequal-mass shape sphere depicted on the right in Fig. 20 I will argue, without as yet considering the beginnings of Newtonian solutions that will be added to it, that you see the essence of cosmic topography: variety, delineated by complexity contours put where they are by universal gravitation. Unlike the equal-mass case, there is no symmetry. That is a potential virtue, not a vice. Effects need causes; they can often be found in mathematics. In our universe we see symmetry in the large but not in the small. The unequal-mass shape sphere is lopsided because it has only three particles; it is a world with one mountain and three bottomless pits. The shape space for a large number of particles with masses that differ by not too much is an extended plateau, a landscape with many smooth-topped elevations (one at each central

---

* *Eigen* is a German word with more than one translation into English; the best in the present case is 'proper'.

configuration) with only slight differences in their shape-potential altitudes and, scattered between them, narrow and infinitely deep wells. Each elevation has its own height, but just one is ever so slightly higher than the rest. Let me call it Alpha.* It is the *primus inter pares*. If we seek the cause of the ultimate effect, the form of the universe we see around us, where can it be if not at the top of the highest elevation? To bring in the quantum question, I think it might even be that from the vantage point Alpha gives us, we see that the road to quantum gravity goes through the Newtonian three-body problem. By that I mean it could give hints of the way through the difficulties that the unification of general relativity and quantum mechanics presents.

It is worth pointing out here that quantum mechanics, which still has baffling elements, could never have been found had it not been for the mutual support that theory and experiment gave each other. In contrast, the search for quantum gravity is almost entirely bereft of experimental support. In its absence, theoreticians can only fall back on whatever principles seem sound and come to hand.** I think that there are two principles that are better than the proverbial search for the house key under a streetlight, and both principles are due to Leibniz; he called them his two great principles. The first asserts the *identity of indiscernibles*, in accordance with which two things with *all* of their attributes identical are one and the same thing. Thus the differently placed triangles in Fig. 6 are, considered as possible shapes of the universe, identical; they would not be if they were simply parts of the universe, because

---

*In *The End of Time* I gave the name Alpha to the point with zero size in the configuration space of the universe. What at that time had not occurred to me was whether a shape could or should be associated with it. However, within a year of the book's publication I had the intuition it must have a shape and moreover one more uniform that any other possible shape. It is only after nearly two decades and with critical contributions from collaborators and information gleaned from N-body specialists that the new notion of Alpha and its significance has crystallised.

**In the current impasse in which they find themselves, most string theorists seem to be reduced to a single principle: there must be no mathematical inconsistencies in the theory. The two principles I am going to propose and the context of their application are very different. As regards string theory itself, about which I have no detailed knowledge, I do think its aim of finding a unification of all the forces of nature is appealing.

the further attributes that define their relation to the rest of the universe would then distinguish them. The first principle, applied to different triangles taken to represent possible shapes of the universe in their own right, dictates that best matching (or an equivalent background-independent procedure) be used to evaluate their difference. Things that differ only in the way we choose to represent them should not have different effects in the world.

Leibniz's other great principle is that there must always be a *sufficient reason* for any real effect. Let's see how it can help in the search for what one might call a comprehensible law of the universe. Newtonian theory allows countless solutions that, projected to shape space, become curves that wander all over its vasty landscape. Why one curve rather than another? There is a possible reason: the law of the universe dictates that their source must be at Alpha. That is the hypothesis to be explored in this chapter and Chapter 17. In Newtonian theory with large N and all the many summits barely below the maximum, singling out Alpha may seem an affront to 'dynamical democracy'. I will do it only pro tem for its illustrative value. There is also the problem that nothing fixes N—it has no upper bound. The hope is that both problems find their resolution in general relativity.

At least the three-body problem, my conjectured source of hints of the way to a classical law of the universe and quantum gravity, has, as the first nontrivial problem in dynamics, a lot going for it when considered from a Leibnizian perspective. Its Alpha, at the equilateral triangle, ranges high over the landscape that it surveys; we will see that very special solutions emanate from the total coincidence at it and descend to explore the foothills of the shape sphere. Like Alpha, these solutions have a Leibnizian pedigree. There is a sufficient reason for their point of origin: central configurations are the only possible loci of total explosions. Their shapes, the only uniquely distinguished points in shape space, are completely determined by the mass ratios of the particles. Since the mass ratios determine not only the position of central configurations but the entire topography that the shape potential lays out on shape space, they also determine the eigendirections of the Hessian at each central configuration. If the law of the universe, judged by its predictive power, is to express as much perfection as possible, Alpha is surely the point of greatest interest. And so the principle of sufficient reason takes us at least this far in our search for the law of the universe. Boltzmann said it would be impossible to explain why we experience phenomena subject to laws. It is hard to argue with that, but a ranking of conceivable laws does seem to be possible. Still, as we will now see, one great mystery remains.

Total Explosions

To explain what this mystery is, I must now describe in the necessary detail nonhomothetic total explosions and why homothetic ones miss exactly half of what is remarkable about total explosions. The half that the homothetic ones do get right is that they begin at a central configuration. As I have emphasised, that is a very special shape of the universe. However, in a homothetic total explosion the shape does not change. Since we insist only change of shape is observable and can be said to be physical, in a homothetic total collision nothing happens. As I already said, it's a fiction—a misconception that survives from Newton's absolute space and time. It's a single shape, not the birth of history. Nonhomothetic total explosions are quite different. They correspond to curves that begin at a central configuration and then go off through shape space. In the rare cases in which the central configuration that is their point of departure is exactly symmetric and the corresponding Hessian has no distinguished eigendirections, the curves can leave in all directions. In contrast, something very different happens in the generic asymmetric case in which the Hessian does have distinct eigendirections with a range of eigenvalues. Let's consider the most interesting central configuration, the one at the absolute maximum of the shape potential. That is simultaneously the absolute minimum of the complexity and thus the most uniform central configuration our model universe can have. The eigenvalues of the Hessian will all be different and can be ordered by their magnitude, beginning with the largest and going down to the smallest.

Now comes the second very special fact: the overwhelming majority of the curves that trace out in shape space the solutions which begin as total explosions leave the central configuration initially *exactly* along one of the back-to-back directions defined by the eigendirection that has the eigenvalue with the largest magnitude; for brevity, let me call it the maximal eigendirection. (I will spell out what 'exactly' means after I have completed the overall description of all possible total collisions.) Curves that start along the maximal eigendirection can then bend away from it in all the remaining directions; for a system with N particles and thus a shape space of $3N - 7$ dimensions, the space of remaining directions has $3N - 8$ dimensions. For a clear image of what can happen, consider a curve on a piece of paper, which has two dimensions; from an initially straight section, the curve can bend away on either side into the orthogonal direction, doing so at any one of infinitely many different rates. The technical expression for this is that there is an infinite one-parameter family of curves. There is a much greater freedom of bending for a wire in three-dimensional space with an initially straight section; it can bend by different amounts in the two directions orthogonal to the wire. There is now a doubling, to a two-parameter family, in the infinity of possible ways to bend. Mathematicians say the one-parameter family forms a set of measure zero within the two-parameter family. Each increase in dimension brings an infinite increase in the number of bending possibilities; more relevant for the present discussion is what happens when the number of possible dimensions is reduced. There is then an infinite decrease in the number of bending possibilities.

This brings me to the remaining total-explosion solutions in the case of N particles, which must be three or more in number. Like the great bulk of curves that leave in one of the back-to-back possibilities along the maximal eigendirection, the curves of these remaining solutions are forced to start along one of the eigendirections. There are many fewer such solutions because of the simple fact that their freedom to bend away from the initial direction is greatly reduced. Thus, for the solutions that leave along the eigendirection with eigenvalue next in magnitude after the maximal one, the number of dimensions into which bending can take place is reduced from $3N - 8$ to $3N - 9$. As the previous discussion shows, this is not a decrease of one, from nine to eight, but a decrease by an infinite number because of the infinite decrease in bending possibilities. There is a further infinite decrease in the number as we proceed down the sequence of eigenvalue magnitudes. When we get to the last but one, there is still an infinite one-parameter family of bending possibilities. However, for the very smallest eigenvalue, just two solutions set off, back to back, along the corresponding eigendirection.

BEFORE I ATTEMPT to explain, as best I can, why the total-explosion solutions are forced to have this remarkable form, I need to say something more about the bending, which is just as remarkable. Let's start in two dimensions with a curve on a sheet of paper; the curve initially is straight but then bends away parabolically from its straight continuation. This means that from the initial point at which the deviation commences, the distance from the straight line increases as the square of the distance from the initial point of departure; there is only an infinitesimal deviation at the start, but it then increases more and more rapidly. If you look at the left of Fig. 2 (page 96) and assume the time axis is lifted so that it just touches the bottom of the parabolic curve, you will see how such an increase happens. There can be an infinite family of such parabolas nested within each other, each bending away from the horizontal by a steadily increasing amount. That makes a one-parameter family of them.

For readers with knowledge of the infinitesimal calculus, the first derivative, which measures the slope of the curve, vanishes at the initial point of bending—at that point the curve is flat and the information about the rate of upward curving is in the second derivative. However, there can be curves for which not only the first derivative but also the second derivative vanishes at the initial point; indeed, finitely many of the derivatives may vanish, in which case the curve stays close to the horizontal axis out to an increasingly great distance. Finally, there is the truly extreme possibility in which all derivatives vanish at the initial point even though there is still a one-parameter family of possibilities of bending away; moreover, there can be a one-parameter family of these possibilities for each dimension into which bending can occur. In such a situation, which is precisely what is realised in total collisions, the curves have an *essential singularity* at the initial point of bending.

By now I hope you have grasped how very special total explosions and their time-reversed interpretation as total collisions are. In starting to give you some idea why they have their remarkable properties, I first want to emphasise that the cause of every effect we are considering is to be found in the topography of shape space. This, as I hope I have made clear, is defined by the contours laid down on it by equal values of the shape potential. These, in turn, are completely determined by the mass ratios of the particles. They determine the entire topography of shape space and in particular the most special location within it—the maximum of the shape potential and, at it, the all-important eigendirections and associated eigenvalues.

Now, how is it that they can exert what looks like an extraordinary controlling effect? I think the best explanation is by analogy with the apple that William Tell shot from the head of his son, which I mentioned in Chapter 11. All would be well if he hit the apple; because it had a finite size, he did not have to get it exactly in the centre. But he did have to get two things right with the tiniest of errors: the aim and the amount by which he drew the bow. The analogy with total collisions is this: how can an incoming solution be 'aimed' so that it 'hits' the central configuration, a mere point in shape space, and, moreover, does so with size (measured by the root-mean-square length ℓ_{rms}) that is exactly zero? In the case of total collisions, there cannot be even the slightest of errors. Among all the possible solutions in shape space, only a very few come anywhere near the central configuration. If they do get close enough and their volume happens to be decreasing toward zero at the proper rate, they can be described in the immediate vicinity of the central configuration in what is called the linear approximation, which is analogous to describing (as in the earlier discussion) the behaviour of a curve at a particular point on it in terms of its direction at that point and its bending at that point. This gives equations that are much easier to handle and can be solved exactly.

In this linear region in which we are considering solution curves that might lead to a total collision, each candidate will have definite proportions of its direction along the various eigendirections. This is just as with compass directions. You can go due north, due east, or by a certain amount east of north, in which case you have components of your direction both north and east. There will in general be components of the direction of any solution we are considering along all of the eigendirections; only in very exceptional (zero-measure) cases will any one of the components be exactly zero. Now, the physical significance of the eigendirections and their eigenvalues is that they identify the directions along which the forces, with strengths proportional to the eigenvalues, act to change the shape of the *N*-body universe in the vicinity of the central configuration. Precisely at it there are no shape-changing forces. It is only away from it, in its immediate vicinity, that these forces begin to be present with strengths and directions as just described.

This means that if in the linear region the solution has a nonvanishing component along any of the eigendirections there will be a tendency for the solution to be pulled in that direction. The remarkable thing which the actual calculations show is that if there is any component along the maximal eigendirection, no matter how small, then willy-nilly the solution will be bent round and forced to come to the central configuration along that direction—and, moreover, in the very remarkable way that makes all the derivatives that characterise the bending become exactly zero at the total collision. Except for the fact that it came in along the maximal eigendirection, all the information about the direction of approach to it—the manner of bending away when we think in terms of total explosions—is totally annihilated. If exceptionally the component of any candidate solution along the maximal eigendirection is exactly zero, then the one with the next-highest eigenvalue becomes dominant and all such solutions must come in along that direction. There are far fewer of these solutions, only a zero-measure

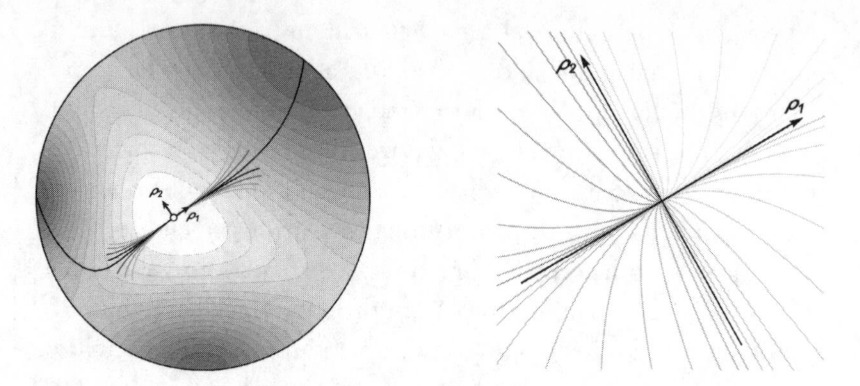

FIGURE 21. Unequal-mass shape sphere with curve of steepest descent and Newtonian solutions (shown in detail on the right) splaying away from the eigendirection with the larger eigenvalue.

set, compared with those of the maximal direction. As already indicated, this drastic step-by-step reduction continues all the way down to the direction with the smallest eigenvalue, which has a solitary associated back-to-back solution pair.

This behaviour is illustrated in Fig. 21 for the three-body problem with the mass ratios of the shape sphere illustrated on the right in Fig. 20; in this three-body case there are just two eigendirections. On the left the eigendirection with the larger eigenvalue, identified by ρ_1, is shown effectively tangent at the summit to the curve (which is not a solution) that from there takes the steepest descent down through the contours of the shape potential. It goes all the way to the shape sphere equator, at one end to the coincidence of the two heaviest particles—the nonhomothetic solution in Fig. 19 is shown heading in that direction with the 1/2 and 1/3 particles bending towards each other—and at the other to the coincidence of the heaviest and lightest particles. Solutions, shown schematically, splay away from the eigendirection on both sides. The eigendirection with the smaller eigenvalue is also shown; it has just the two solutions that go off in opposite directions along it.

The curves on the right in Fig. 21 are the result of actual calculations of total-explosion solution curves in the immediate vicinity of the point of total coincidence. The eigendirection with the larger eigenvalue is identified by ρ_1, the one with smaller eigenvalue by ρ_2. You can see how all the curves bend round to come in along the ρ_1 eigendirection. The bending at the very end of the curves that initially are virtually parallel to the ρ_2 eigendirection is very pronounced. The curves are shown for the solutions that can leave (or, with time direction reversed, approach) the central configuration in both of the ρ_1 back-to-back directions (on the two sides of the ρ_2 eigendirection). This illustrates the difficulty *N*-body specialists have in proposing any way to link up solutions on the two sides of the central configuration. This is because from both sides all solutions come in exactly along the eigendirection, which means that any solution on one side can be joined smoothly to any solution on the other side.

An important thing to note here is the way in which the given state of affairs is to be described as a total explosion. Initially, on the immediate departure from the total coincidence, the maximal eigendirection is in complete control of the motion. However, after a while the eigendirections with smaller eigenvalues begin to have their say and start to pull the solutions in their direction. This can be clearly seen on the right of Fig. 21. The amount corresponds to the proportions of the motion in the linear regime that are along the various eigendirections. Nothing in the theory allows us to predict them. We need to fall back on statistical arguments. Whatever the proportions are, the outcome will be a solution that eventually settles down and sets off on its way through shape space.

At the end of Chapter 1 I said that two kinds of Janus points can exist. By establishing a comparison with the examples of the first kind considered up to now, we now know what those of the second kind look like. The difference is substantial. Janus has undergone a triple change. He is no longer at an arbitrary point in shape space; he now stands on the summit of the shape potential. That's the most significant change. Next, he is not a back-to-back god looking out in opposite directions onto a single solution that passes continuously through him. In the book up to now his locus on any solution was special, but it was not a terminus for either half-solution; they passed smoothly and without ambiguity through him. Now there are whole families of half-solutions on each side, splaying out from each of the eigendirections; that's the second change. The third change has just been stated: mathematical research has failed to find any way in which a solution curve that reaches Janus in a total collision from one side can be continued in a uniquely determined way through to his other side. The only mathematically well-defined continuation that exists joins the two halves of each eigendirection. The Newtonian trajectories that peel off from the two faces of Janus cannot be linked together. However, time-reversal symmetry is still respected for the initial directions and for the two sets of infinitely many half-solutions. As Boltzmann required, the two directions of time have 'equal rights', though not in the way he imagined. Janus is a god of beginnings, not just of one on each side but of many.

A PARTICULARLY INTERESTING thing that Fig. 21 illustrates is the emergence of what looks like a complete set of normal initial data, with no restriction on the freedom to specify both positions and momenta even though at the start there is no freedom at all apart from the two opposite ways to exit from total coincidence. Except for the Janus-like back-to-back ambiguity, all solutions have exactly the same initial shape and direction of motion. There are just two families of solutions, each completely constrained on their respective sides. In an infinitesimal neighbourhood of the total coincidence at the central configuration there is no information at all about the directions that the solutions will then take. All the information-bearing derivatives are exactly zero. However, at any finite distance (no matter how small) all possible solution directions are present, though their emergence from one single shape is still very obvious. Even that striking fact ceases to be immediately manifest with increasing distance from the central configuration. This is because the solutions can twist and turn in many different directions as they take their course through shape space. Superficial examination of the solution curves over a small region would give no indication of their very special origin. Precise knowledge of the solution curve, sufficient to track its course back to its origin, would be needed to establish its birth conditions. In the real universe, which much evidence shows is governed to great accuracy by Einstein's general theory of relativity and will be the subject of Chapters 17 and 18, light, with its finite speed of propagation, can make a big difference. It allows us to look back toward the universe's birth, if not to the very moment then to a point at least close to it, and see directly evidence of its nature.

Even without light and all the information it bears, Fig. 21 also highlights the mystery I mentioned at the end of the previous part of this chapter. It concerns the amount by which, under the influence of the remaining eigendirections, the various curves in Fig. 21 bend away from the eigendirection along which they all begin. The resultant emergent value is fixed at any finite distance from the point of total explosion and after that evolves in a deterministic manner. In the final part of Chapter 8 I called it the creation measure since it measures the proportion of the total kinetic energy that is in change of shape and therefore changing—indeed, increasing—the amount of structure in the universe. The complete set of solutions in the figure and the analogous solution sets for any number of particles are completely determined by the topography of shape space, which in turn is determined by the mass ratios and the formula for the shape potential. The mystery is this: what cause, if any, determines the particular universe, each with its own creation measure, that is realised? It is an element of Newtonian theory quite unrelated to mass ratios. It seems that a lottery brings a universe into contingent existence; why it and not some other is inscrutable. The Newtonian solutions in shape space are the realm of possible universes. If we seek the birthright of a chosen one, we will not find it.

It must be born at the point of total coincidence; what is effectively the toss of a coin sends it from there in one or the other of the back-to-back directions. But then the creation measure takes over. It is like a breeze that inclines a child making her first steps to go one way rather than another. The choice once made, at least within classical physics, is final but, to say the least, enigmatic. The created universe is an effect without a cause. Leibniz would turn in his Hanoverian grave and cry out that the universe begins in the state of greatest possible uniformity, the source, and evolves from it to a destiny of the greatest possible variety. The source is not the problem; there is a unique shape that has the greatest uniformity and, remarkably, back-to-back directions from it. Each can set a universe on its way. But then a breeze makes the choice that counts. If Platonic being is shape space, endowed as it is by mathematics with Alpha and the contours on its landscape that prescribe directions, the breeze is becoming.

A thing is ineffable if it cannot be expressed in words. By analogy, the cause of the creation measure is 'indepictable': in contrast to the contours of the shape potential, no topographical feature in shape space determines its value, that is, the direction in which it will choose to send the evolution curve of the universe and create history: becoming within being. Is there really no reason why I am writing this now and you are reading it?

Perhaps there is, but it can hardly be found within conventional science, or at least not as it now exists. Does the creation measure heed the posthumous call from Hanover and send the universe on the course that will, in the fullness of time, create maximal variety? Most scientists will immediately object to that, and with good reason. Ever since Darwin first presented his highly successful theory of evolution, the notion that beneficent intelligent design explains what is found in the world has been discredited. The consensus today is that science can seek answers only to questions of how, not why. I will merely observe that the description of total explosions presented here seems to have whittled everything down to a picture in which all manifest causes for our existence have been identified but not why one history of the universe rather than another is realised. We seem to be left with some form of the idea that Boltzmann's old assistant Dr Schuetz first proposed: if it were not for the almost inordinate amount of structure that has been created around us in the universe, we would not be here to see it.

Besides its intrinsic interest, one of the main reasons I have included in this chapter more or less full details of Newtonian total explosions is, as I already said, because of the hints that they might give of what the big bang in Einstein's theory is like. In this connection I hope you recall what Schrödinger said about the sovereign status of the number zero. Shakespeare wrote the play *Much Ado About Nothing*; this chapter too might be said to be about nothing, or zero. In fact, the chapter has completed a hierarchy of zeros. The first, the vanishing slope of the square of the root-mean-square length ℓ_{rms} at the Janus point in Fig. 2 (page 96), which can be anywhere in shape space, has dominated the story up to now. By itself it explains time's arrows. A consequence of the Janus point's special nature, not hitherto mentioned, is that at this point the equations which govern the system's trajectory in shape space become simpler; it is the unique point at which they are more nearly straight than anywhere else. That's the effect of the first zero. The second zero appears if, without ℓ_{rms} becoming zero, the Janus point is situated at an extremum of the shape potential—for example, at the equilateral triangle in the three-body problem. At such a Janus point, the shape potential is flat—its gradient (slope) is zero in all directions. It is this zero that directly reflects physical reality. Indeed, since it is the gradient of the shape potential which causes the Newtonian trajectory to bend, the form of the system's trajectory in shape space becomes even simpler—approximating straight behaviour over a greater extent—than for a Janus point anywhere in shape space. It is when we come to the third zero—the vanishing of ℓ_{rms} at the Janus point—that something truly remarkable happens: a total collision. In fact, the simple vanishing of size that characterises a total collision is alone so powerful that it determines not only the locus of the cataclysmic event but also the remarkably specific manner in which it happens. Whereas the first zero enforced a certain straightening of the curve of history and the second enhanced the effect, the third leads to an essential singularity with total straightening of the curve at the point of zero size. If that were not enough, the angular momentum must also vanish. If the Newtonian big bang is so restrictive, what does the zero size at the Einsteinian big bang enforce on the shape and behaviour of the universe?

Statistical Implications of the Royal Zero

Whether, with me, you conjecture that a total explosion at Alpha characterises the law of the universe—a further argument for it will be advanced in Chapter 17—or you conceive it as something that, in the form of a past hypothesis, it is necessary to add to the physical laws, a total explosion certainly has a dramatic effect. It has the virtue of showing what the effect of a past hypothesis, if it were one, might actually be. It is very striking. Any possible history of the universe must begin at Alpha, the preeminent location in the topography of shape space, and set off along one of the eigendirections that the immediately ambient topography of Alpha uniquely determines. Not only that, the initial velocities of the particles are completely determined and strong statistical predictions can be made about how they will be changed by the influence of the eigendirections with smaller eigenvalues than the maximal one. A remarkably predictive theory is obtained. As Boltzmann hesitantly but Feynman and Penrose more emphatically suggested, our model of the universe does have an exceptional origin. The question now arises of whether and how we can think about it in entropic terms.

One thing is sure: the envisaged scenario allows far fewer microhistories to exist. Compared with what general Newtonian theory allows, they form a set of measure zero. Any probability one might try to define for them is zero. The reason for this is that the initial shape of the system is completely fixed, whereas normally there is no restriction. Thus all possible configurational entropy is completely wiped out. Although the solutions that escape from the initial straitjacket of their eigendirection do develop an effectively complete set of momenta and a corresponding contribution to the number of microhistories, the overall reduction is still huge. What stands out above all is not only the unique initial shape but its uniformity, as illustrated in Fig. 18 (page 212).

Despite the huge reduction in the number of microhistories, and with it the associated entropy, I think something like a conventional entropy associated with ignorance could arise quite early in the evolution from Alpha. The central-configuration shape of the

500-particle system with shape potential very near its maximum of Fig. 18 does resemble the swarm of bees invoked as the origin of the 'most beautiful system'. As I will point out in Chapter 18, it is also strikingly like the distribution on the right of Fig. 11 (page 142); I noted in Chapter 14 that such distributions occupy a tiny fraction of shape space. Moreover, as I argued earlier, in any one of the solutions that sets off along one of the back-to-back eigendirections—almost certainly the one with the maximal eigenvalue—the initial velocities will, despite being completely fixed, almost certainly have something close to a Maxwellian distribution of their kinetic energies as they emerge from an initially almost perfectly spherically symmetric collection of particles. Note also that the overall expansion in the extended representation, the expansion of the universe in the conventional account, is invisible for internal observers. All that they have any chance of seeing initially is the change of shape of the universe along the maximal eigendirection.

It is only bit by bit, as the tight control of the eigendirection is relaxed, that gravitational interactions between neighbouring particles come into play. It will be the amount of bending of the solution curve in shape space as it emerges from total coincidence that determines, through the ratio of the shape-changing and expansive (dilatational) kinetic energies, how strongly these interactions are manifested. Thus, the change in the shape of the universe from an initial one like that of Fig. 18 proceeds in two characteristically different ways: at first holistically along the eigendirection and then parochially through interactions of particles with their nearest neighbours. They can exchange energy with each other and begin to form clusters, especially Kepler pairs. This will transform a particle distribution of the kind shown on the right of Fig. 11 into one of the kind on the left. The picture presented in Chapters 12, 14, and 15 will, suitably modified, still hold. The velocity distribution assumed in those chapters for the state at the Janus point will be shifted to some distance from it. The statistical arguments of the earlier chapters should still hold at the displaced position.

In this new situation a better metaphor than an infinite dartboard and a blindfolded creator is a dartboard—shape space—of finite size and a creator without a blindfold but so shortsighted that he can see only roughly where the bullseye is. He will never hit Alpha but has a good chance of landing darts in the region where there is something like thermal motion. In it the huge reduction in the number of microhistories down from standard Newtonian theory to those the creation measure is free to 'bring to life' as they leave Alpha will be hidden to a considerable degree. This is because once interactions between the particles have disrupted the original highly ordered initial shape and motions the current state will look much like one that might be expected in any Newtonian solution in which there is overall expansion. The only evidence for something special in the past may be residual traces of the shape of the central configuration and special eigendirection at Alpha; this possibility, should general relativity have an analogous Alpha, will be considered in Chapter 18. Of course, there will also be the pervasive temporal asymmetry of phenomena observed far and wide. However, there will be little sign of the 'conspiracy' that becomes manifest if the motions are run backwards in time to total collision.

Feynman did not call for a truly extraordinary early state of the universe, only for one that "had a very low entropy for its energy content". In contrast, Roger Penrose, pondering the origin of the second law of thermodynamics after the discovery that black holes have a colossal entropy, thinks that the universe's early state was indeed extraordinary. Many of my readers will be familiar with Penrose's graphic image in which he inverts the needle-in-a-haystack simile and depicts the divinity sticking a pin into a haystack. The phase space of the universe is the haystack and the divinity's hope is that the pinhead will finish up at a point of inconceivably low entropy that ever after can increase in accordance with the second law. Penrose estimates the divinity's chance of success at less than one in 10^{121}.

Now, shape space is not like the largely undifferentiated phase spaces of confined dynamical systems. It is not a haystack. It's a country whose topography boasts many summits, one of which is the highest. Unlike the shortsighted creator invoked just now, a true divinity surely has good eyesight. He simply has to stick the pin at Alpha. This will automatically ensure that the overwhelming majority of solutions begin their life along the dominant eigendirection. The creation measure can then choose among all the allowed histories of the universe that can branch out from it. As I will argue in Chapter 19, the second law of thermodynamics, suitably defined for effectively confined systems that form far from the Janus point, will hold in all of them. However, the reason for this will not be the special initial condition itself but rather the expansion of the universe that happens on either side of a Janus point whether or not the size of the universe is zero or finite at it. As long as clumping, due predominantly to gravity but also to other forces, is allowed by the expansion to take place, the complexity of the universe will increase and that will determine the direction of time.

Of course, if the divinity does not have the expected acuity of vision, the pin might be put not at the summit of Alpha but, when there are many particles in the universe, in one of the great number of summits with almost the same altitude. When we come to general relativity in Chapter 17, we will find a tentative hint that, just as in the three-body problem, there is a solitary mountain. There may not be any rivals.

Hints of Quantum Mechanics

I have written this part of the chapter in the hope that readers will have some knowledge of basic quantum facts from one of the many books for the lay reader that now exist. To include more about quantum issues than I do in what follows would increase the already rather ample length of this chapter. However, I do want to draw attention to an uncanny parallel between the most distinctive manner in which quantum and classical mechanics differ and what we find in zero-energy total-explosion solutions of the Newtonian *N*-body problem. In all of normal classical dynamics the positions and momenta of the particles are simultaneously observable and can be freely specified as initial data. In Machian dynamics that freedom is significantly restricted. Both the angular momentum and the energy must be zero. The restriction, for which sound relational arguments can be given, is a gain, not a loss, because with it Newtonian theory becomes more predictive. The elimination of energy and angular momentum together with the representation in shape space eliminates all inessentials and also reveals clearly the way the creation measure takes over from time as the dominant player in the dynamics of the universe. The effect it has is the opposite of entropy's. It does not disrupt; it creates.

It is when the total-explosion condition is added that, as we have seen, there is a dramatic change; it is where the quantum parallel appears. In normal classical dynamics there is, as just said, complete freedom to specify initial positions and momenta; in contrast, in the scenario we have been considering, that freedom is drastically reduced, with the number of degrees of freedom essentially halved because the initial positions must be those of the central configuration. This is very like (although not identical to) what happens in quantum mechanics. According to the famous uncertainty principle of Heisenberg, it is quite impossible to have simultaneous precise knowledge of both the position and the momentum of a particle. The more accurately one is known, the less accurate is the information about the other. There is a similar reduction when many particles are considered; as regards what can be known about them, there is an exact halving compared with the situation in classical physics. Mystery still surrounds the manner in which the macroscopic world we find around us, which behaves classically at least to a high degree, emerges from quantum uncertainty. I will say something about that shortly; here I merely wish to emphasise the parallel between the quantum halving in the number of observable degrees of freedom and what we have seen in total explosions. I want to consider what its implications might be for finding the grail of physics: a consistent theory of quantum gravity and, with it, a quantum theory of the universe.

Quantum theory was found by a process called quantisation. In it, the starting point is a classical theory. In fact, the very first such theory, used by Heisenberg in the breakthrough to quantum mechanics in the summer of 1925, was what is called a harmonic oscillator. An example of one is a pendulum in the situation in which the swings it makes are not too large. As discussed earlier in the book, the period of the pendulum does not depend on the swing amplitude as long as it remains small. Any system which behaves like that is called a harmonic oscillator; it's the simplest nontrivial dynamical system one can have. What Heisenberg did was to create a parallel mathematical formalism in which the fundamental classical concepts of position and momentum were replaced by mathematical entities called operators. I will not attempt to describe them, how they are used, nor how they relate to actual experiments. However, I do want to point out that, to this day, quantum mechanics has by no means broken free of some kind of classical background.

I hope you will recall from Chapter 13 how, in her famous theorem, Emmy Noether proved the existence of two quite different kinds of classical dynamical theories, in one of which (described in part I of the theorem) there exist nonvanishing conserved quantities like energy, momentum, and angular momentum, while for the theories described in part II, as shown in Dirac's subsequent treatment, there are only combinations of the momenta that must vanish exactly. Noether did not attempt to explain why the two kinds of theory exist or whether there might be some relationship between them. She simply presented a mathematical framework in which the two kinds of theory exist. In that chapter, I pointed out that in the relational Janus-point framework it is perfectly possible to understand how systems described by part I of her theorem, which require a uniform spatiotemporal background, emerge as increasingly well isolated subsystems of a universe governed by a law of the kind described in part II of the theorem. This showed how Newton's framework of absolute space and time, reexpressed in modern terms as an inertial frame of reference, can be shown to be in fact the manifestation of the all-powerful control of the universe at large.

The important thing about this is that the quantisation process, for all the brilliant successes it has achieved, has always worked within what is essentially the classical framework of an inertial frame of reference, including what is effectively an absolute time. Right from the birth of quantum mechanics, the special role of time was noted—it remained classical and did not become an operator. Until the appearance of the many-worlds interpretation of quantum mechanics (which remains controversial and which I will discuss shortly), it also proved impossible to formulate and interpret quantum mechanics without the classical background of macroscopic bodies governed by dynamical laws defined in an inertial frame of reference. At the pragmatic level, this is the basis of what is known as the Copenhagen interpretation of quantum mechanics, which Heisenberg and Niels Bohr formulated when Heisenberg visited Bohr in Copenhagen shortly after the discovery of quantum mechanics. For the purposes of the present discussion, the point I want to make is that, to this day, all successful quantisation has treated positions and momenta on an equal footing in a classical background whose existence has not been explained but, if not taken for granted, at least could not be dispensed with.

What clues might we gain from the background-independent shape-dynamic representation of the universe presented in this book, especially if we posit that the quantum theory of the universe must use at least some classical concepts and reflect a total-collision birth of the universe? I think the first change must be to stop treating positions and momenta on an equal footing. The fact is that a momentum is a direction *and* a magnitude, the speed associated with the direction. Now, in normal dynamical theories this speed is defined relative to an external time. In contrast, in shape dynamics the speed of an individual body is simply a measure of how large its change of position is relative to the changes in position of all the other bodies in the universe. There is no background independent time. This means the magnitude which has to be added to a direction to make a momentum is suspect. Moreover, all the changes of position are changes of bodies relative to each other. Added to this inescapable fact is another: all observations are ultimately of relative positions. When we measure our temperature with a mercury thermometer, we look to see the position of the head of the mercury column against the marks on the glass tube. Even information expressed through printed words on paper relies on the relative positions of different amounts of ink on the page. John Bell, whose work revealed some of the most profound mysteries of quantum mechanics (and which will be discussed shortly), always insisted on the primacy of positions.

If, as I have argued, the history of a classical universe is nothing more and nothing less than a succession of shapes from which the notion of duration emerges, I think it is clear that the attempt to create a quantum theory of the universe should certainly start without the notion of a preexisting external time. In fact, the creators of one of the main approaches to quantum gravity, so-called canonical quantisation (which basically followed Heisenberg's pioneering paper), were forced to consider such an option by the dynamical structure of general relativity. In its most consistent application, it led to the complete disappearance of time and a static quantum theory of the universe. Various suggestions for how the evidence for (or at least appearance of) change and the passage of time could be recovered have been made; my own *The End of Time* was one. I will return to that issue in the final chapter.

There is one aspect of quantum theory that I do think will be explainable with the notion of the Janus point, even in the more modest form of that notion considered in the chapters prior to this one. I am referring to the arrow of time traditionally associated with the notorious collapse of the wave function. In 1957 Hugh Everett III proposed an explanation of this mysterious phenomenon that has become known as the many-worlds interpretation of quantum mechanics. According to it, wave functions do not collapse but instead split into many branches, taking the observer who makes the measurement with the one he or she has observed. Some quantum theoreticians go so far as to suggest that any observation of the universe is collapsing its wave function and is putting not only the observer but also the entire universe into one among many possible states. One of the issues related to the many-worlds interpretation is whether, when it is applied to subsystems of the universe, there is not only branching of their wave functions forwards in time but also, in accordance with the time-reversal symmetry that quantum theory shares with classical physics, branching backwards in time. I feel rather confident that is not the case, first because I think it is a mistake to try to separate out parts of the universe as if they were dynamically isolated systems and second because the 'role-reversal' argument, which takes on a particularly strong form in total-explosion solutions, has the potential to provide a comprehensive explanation of all arrows of time.

More speculatively, I now want to consider possible implications of total explosions for another of the deep mysteries of quantum mechanics: the effect known as *entanglement* (the original German term, *Verschränkung*, was coined by Schrödinger). It is an effect considerably more puzzling than the fate of his metaphorical cat, whose wave function allows it to be both dead and alive until an observation is made. It was John Bell who uncovered the full extent of the entanglement mystery. It is all to do with seemingly inexplicable correlations expressed as inequalities that now carry his name. Classical dynamics has plenty of correlations that are explicable. The three-body solutions of the minimal model in Chapter 7 all have vanishing energy. Once the particles have separated into the singleton and Kepler pair, the singleton has positive energy while the Kepler pair's is negative. If you measure the singleton's energy, you immediately know the Kepler pair's. Interconnections like this entered physics through Newton's creation of dynamics; they are already a wonder, but the Bell inequalities are much more remarkable. Experimentalists can create pairs of particles with correlated spins—quanta of angular momentum that sum to zero in a so-called singlet state—that then fly as far apart as desired before measurement of their spins is made. Already in 1935 Einstein, in a paper known by the acronym EPR (E for Einstein, P for coauthor Boris Podolsky, and R for coauthor Nathan Rosen), had shown the remarkable degree of correlation that can be established in an analogous situation in which position and momentum rather than spins are measured independently at spatially separated locations.

It was the EPR paper that prompted Schrödinger to coin the term 'entanglement' and illustrate a feature of it with his dead-and-alive cat. But neither he nor EPR plumbed the full depths of the quantum mystery. It was Bell who, in 1964, did that by exploring the implications of measurements made independently with spin-detecting instruments inclined arbitrarily—that was the critical insight—at the two different locations where the spins are detected. He showed that within the framework of standard quantum theory the degree of correlation between the measured spins that quantum mechanics predicts could not be explained by any classical theory unless nature employs 'spooky action at a distance' communicated at speeds exceeding that of light (and therefore assumed to be impossible in classical physics in accordance with Einstein's theory of relativity). In experiments made between 1980 and 1982, Alain Aspect in Paris confirmed Bell's deduction (from the standard quantum rules) of the hitherto unrecognised correlations. The kinds of correlations he observed in his laboratory were subsequently confirmed by Anton Zeilinger and his collaborators between a mountain peak in the Canary Islands and the Moroccan coast, a distance of more than 100 kilometers. Nobody can currently put a limit on the distance between which such correlations can be observed.

In a BBC radio interview in 1985, Bell commented that superluminal speeds and spooky action at a distance could be avoided if one accepts absolute determinism in the universe and the complete absence of free will. The world would be "superdeterministic, with not just inanimate nature running on behind-the-scenes clockwork, but with our behaviour, including our belief that we are free to choose to do one experiment rather than another, absolutely predetermined". It seems to me possible that the classical picture of total explosions described in this chapter could cast some light on the mystery of entanglement and whether it implies superdeterminism in the universe. The evidence, always assuming it is reliable, is certainly ambiguous. The initial behaviour, right at the total explosion, is surely superdeterministic. However, the creation measure then sets the universe free, or at least it does to a certain degree. In fact, one might even see here a variation on the nature-versus-nurture debate first brought into clear focus in

Prospero's description of Caliban as "a born devil on whose nature nurture can never stick". Indeed, all the individual classical total-explosion solutions are born with the same DNA, so to speak: the central configuration from which they emanate and the eigendirection they are forced to follow. However, each solution is then guided—nurtured—by the structure of shape space, the environment through which it passes. Both Ariel and Caliban, who has the finest verse in the play, are set free at the end of *The Tempest*.

Here one aspect of quantum theory and the classical theory from which it is obtained by quantisation seems to me to be particularly interesting. Under conditions in which interactions play no role in the classical theory, the resulting quantum theory becomes particularly simple. In fact, the wave function that is obtained simply represents, in a well-defined sense, a bundle of simultaneously present classical solutions. This is known as the Wentzel-Kramers-Brillouin (WKB) regime. Since interactions between particles are absent in the initial stages, this suggests that the initial quantum state of the universe will be simultaneously simple but tightly correlated. Whatever form the wave function of the universe might take in its subsequent evolution over shape space, it can hardly shake off the nature of its birth. This cannot be more than speculation, but it does seem to me that the quantum theory of the universe may combine profound unity in the conditions of its birth with unlimited opportunity for subsequent creative development. I can see no limit to the variety of possible shapes of the universe nor any reason why the wave function of the universe should not be free to visit any or all of them. When told by Prospero "Thou shalt ere long be free", Ariel sings:

> Where the bee sucks, there suck I:
> In a cowslip's bell I lie;
> There I couch when owls do cry.
> On the bat's back I do fly
> After summer merrily.
> Merrily, merrily shall I live now
> Under the blossom that hangs on the bough.

The Spontaneity of Artistic Creation

Given the speculative ideas already presented in it, I should probably end the chapter at this point, but it is tempting to ask what the implications of the total-explosion proposal might be for free will and the hard question of how there can be consciousness in a purely material world. Boltzmann said that we could never explain why we experience phenomena which unfold in accordance with definite rules; that surely does remain beyond our ken. As regards free will, Mach said it was an illusion which arises from our frequent experience of wishing to do something and succeeding in doing so. But the origin of the wish itself is inexplicable. We can do what we will, but we cannot will what we will. The desire comes spontaneously. What I have suggested in this chapter as a theory of the universe is already a speculation. I will now hazard speculation piled on speculation, Pelion on Ossa.

The word 'universe' derives from the Latin *universum*, which means 'combined into one'. It's a big if, but if the Alpha idea is in essence correct and, as I just conjectured the quantum theory of the universe combines both profound unity with freedom of creative development, inspiration may come to us at any moment from awareness that our unconscious, being part of it, has of the *universum*. New thoughts might be subliminal inspirations from the whole universe. They will not have been transmitted superluminally in the manner that bothered John Bell but have been nascent in us all the time. After all, as I was at pains to stress in describing the topography of shape space, Alpha is its quintessential expression.

Musicians, who like all of us are tied by uncountable bonds to the universe at large, are vehicles through which rare moments, new but unrepeatable, come into the world. In the epilogue to *The End of Time*, I recount how I had heard Dame Janet Baker interviewed on radio. If some performance of one of the great arias—say, 'What shall I do without thee?' in *Orfeo ed Euridice*—had gone especially well, she was asked, would she attempt to repeat it? Her answer was, "Absolutely not. That would destroy the magic of the Now".

Some years later I heard her again, this time talking about her interaction in opera with the Dutch conductor Bernard Haitink (who, at the time of writing, has just turned ninety-one). She spoke of the way one or the other of them could, on the spur of the moment, be moved to take things in some particular, ever so slightly different way. Musical performance always allows some freedom. Her rapport with Haitink, transmitted between stage and conductor's podium, was so good the other would follow instinctively. Between them something unforgettable would result. What it is about the universe that makes this possible? One can't help wondering.

More recently, I was present at a rather grand dinner at the Fishmonger's Hall overlooking the Thames in London and had an opportunity to talk for a few minutes after the dinner to a famous bass who was also present. As the discussion was 'off the record', I won't name him, but I recounted what Dame Janet had said about working with Haitink and asked whether he had experiences like that. After a pause, he said, "With Haitink, yes, Bernard's special". For a moment the great man seemed to slip into reverie, saying, almost to himself, "Yes, with Bernard over the footlights". Once the inspiration has come, it is clearly maintained at least in part over the footlights. But where does it come from in the first place and take the turns that it does?

In this age of artificial intelligence, it is often wondered whether computer programmes are capable of genuine creation. I have read that the novels they 'write' peter out in triviality. The block seems to be that they can only build on what has already been fed into them, and that is a finite resource. Even the incredibly strong AlphaGo programme that, given the rules, taught itself to play Go to superhuman level operates within a strictly circumscribed domain that is algorithmically defined. Moreover, the hardware of computers is embedded in the universe but in not remotely as sensitive a way as we, flesh and blood, are. Does that forever cut them off from the nuances of the breeze that moves the universe? Shakespeare wrote his sonnets subject to the strictest rules but drew on a whole universe for precisely 140 syllables. Will computers, reacting to the magic and needs of the Now, ever rival them?

CHAPTER 17

THE BIG BANG IN GENERAL RELATIVITY

BY NOW THE READER MAY BE IMPATIENT TO KNOW WHETHER A JANUS POINT, in the form of an Alpha, exists in general relativity. Without one the most ambitious part of this book's project collapses. My collaborators addressed the question in 2015 and answered it in the affirmative, at least in the case of the simplest nontrivial solutions that exist in Einstein's theory. Those solutions share important properties with the particle solutions, but they also have differences that are both interesting and significant, the latter above all because in general relativity big-bang solutions are typical—the technical term is 'generic'—but in Newtonian theory they are extreme rarities. Nevertheless, general relativity does, depending on the matter content, allow many solutions that do not have a big bang. Indeed, many seem to be downright unphysical. The best-known, which troubled Einstein as soon as he became aware of their possible existence a year before he finally found the definitive equations of his theory, involve so-called closed timelike curves. These have given rise to endless debates about the possibility of travelling back into the past and killing your grandfather, a manifest paradox that rules out your own existence.

I have already suggested that such problems in the theory may arise from its creation in the manner laid down by Maxwell, in which differential equations that hold in infinitesimally small spacetime regions are assumed to express the fundamental laws of nature. The alternative Machian view seeks in the first place a law of the universe with an architectonic structure dictated by first principles of the kind Leibniz advocated. Such a law must of course reproduce all the phenomena that hold locally, at least in our epoch. We have seen an example of this in the *N*-body problem in the total-collision solutions. They constitute a zero-measure set among all possible solutions but suffice to describe all locally observed Newtonian solutions—with, moreover, the added benefit of showing exactly how the framework in which they hold arises. The same thing may be true in Einstein's theory. When all inessentials have been stripped from its shape-dynamical core, we will see it naked—Michelangelo's David, all its inner strength revealed.

Compared with Newtonian theory, the difficulties are greater by an order of magnitude. Let's start with the simpler material; it's no good trying to run before we can walk. At the big bang, in accordance with the most typical accounts, the universe came into existence out of nothing in a mighty explosion. At that extraordinary moment, the universe had zero size but an infinite density of matter within it. This at least is what the equations of Einstein's theory, taken literally, suggest. The appearance of infinities in a theory makes physicists uncomfortable; they take it as a sign that the theory breaks down and requires replacement by something better or at least significant modification. In the case of Einstein's theory, it has become a commonplace to say that "general relativity predicts its own demise", which—you may have guessed—is another of John Wheeler's aphorisms. It is applied not only at the big bang but also in black holes, where the theory predicts infinite densities as well. Like total collisions in Newton's theory, these locations in spacetime are singularities; only quantum gravity, it is held, can 'lance the carbuncles'.

I agree with the need to quantise gravity but suggest it may not be as difficult as is widely believed if, as in Newton's theory in Chapter

16, the 'robes' of the extended representation—inertial frames, time, and scale—are removed to reveal the anatomy of Einstein's statue. This matches a process mathematicians call regularisation. In the *N*-body problem we saw that as a total collision is approached, the 'conspiracy' by which the particles contrive to meet all at one place requires the forces with which they act on each other directly to vanish. The shape potential becomes flat and is effectively non-existent. As my collaborators have been saying for some time, this is good news. Quantisation is simplest when no potential is present—the WKB approximation mentioned in Chapter 16 holds.

To clear away the thicket that defends the grail, it is necessary to eliminate extraneous elements. They introduce problems that in reality do not exist. Density is defined as an amount of mass divided by the volume it occupies. That's unproblematic in a laboratory with scales to weigh the mass and rulers to measure the volume. But in extrapolation to the universe, where are the scales and rod? In his *Autobiographical Notes*, published in 1948, seven years before his death, Einstein acknowledged that the conceptual structure of general relativity contained a flaw. Besides the notion of four-dimensional spacetime, it introduced measuring rods and clocks as independent elements. Einstein commented: "This, in a certain sense, is inconsistent; strictly speaking measuring rods and clocks should be represented as solutions of the basic equations (objects consisting of moving atomic configurations), not, as it were, as theoretically self-sufficient entities". He argued that, in the absence of a theory of measuring rods and clocks, his procedure was a justifiable stopgap that allowed physical interpretation of the theory, but the defect, which he called a sin, should be eliminated "at a later stage of the theory".

The Catholic Church distinguishes venial and mortal sins. Einstein's 'sin' was surely venial at the time he created general relativity but almost certainly mortal in black holes and at the big bang. Kepler pairs emerge as rods and clocks from the basic equations of *N*-body theory, but only far from the place where Newton's big bang is to be found in shape space. There are no rods and clocks at Alpha, just the only thing that my grandson Jonah could find in the three-body homothetic total collision: the equilateral triangle. But Jonah would see that the shape-dynamic equations describing nonhomothetic motion do not break down; their solutions actually take a simpler form. For several decades the singularities in Einstein's theory have been an active field of research but without, so far as I know, directly addressing the elimination of Einstein's 'sin'.

TIM KOSLOWSKI, FLAVIO Mercati, and David Sloan took the first steps towards that at the end of 2015; following peer review, their paper 'Through the big bang' appeared in early 2018. It treats the simplest nontrivial model of an Einsteinian universe in which the shape changes. The relational three-body problem has its two mass-weighted shape degrees corresponding to two internal angles of a triangle. They have analogues in Einstein's theory that describe the shape of a space that in three dimensions closes up on itself. The simplest illustration of that in two dimensions is the surface of a soccer ball. Its surface is both homogeneous—the same everywhere—and isotropic, which means that it looks the same in every direction. All planes that cut the sphere's surface and pass through its centre define great circles of the same perimeter. In contrast to a sphere, the surface of a rugby ball, like a zeppelin's, is neither homogeneous nor isotropic. In three dimensions there are geometrical objects called 'squashed three-spheres' that are anisotropic—the distances traversed in going right round them in three orthogonal directions are different—but despite that they are homogeneous. The simultaneous possession of these two attributes makes them diabolically hard to visualise. A zeppelin is surely anisotopic but definitely not homogeneous. The best explanation I can give is in terms of the three local curvatures at each space point in such spaces. Two dimensionless ratios can be formed from their dimensionful values; the homogeneity of squashed three-spheres means these two ratios are the same everywhere. Mathematicians do conceive amazing things. The American Charles Misner used the two dimensionless ratios in a much-cited paper of 1969; they are often called Misner anisotropy variables. They are analogous to the two internal angles that define the shape of a triangle.

One of the beauties of Einstein's theory is that it does not require matter to be present for dynamical evolution to take place. Geometry can change its shape all by itself; this is quite unlike Newton's theory, in which mass is an essential ingredient. Einstein's theory is, moreover, very flexible; it can incorporate many different forms of matter. In fact, Einstein came to see this as a deficiency. His dream, from the early 1920s to his death in 1955, was a theory in which there is no matter at all, just curved geometry that evolves in a way which gives it the appearance of the matter we find around us. He never found the theory.

To make progress we do not need to wait for fulfilment of Einstein's dream. We can make a start with squashed three-sphere solutions; they have an interest that, if anything, is even greater than those of the lopsided unequal-mass shape sphere. They are the simplest nontrivial models of general relativity in which the shape does change and there is an analogue of the shape potential in the particle dynamics. They belong to the class of solutions of the dynamical system called Bianchi IX after the Italian mathematician Luigi Bianchi. In 1898 he identified mathematical structures that are now used to characterise nine possible dynamics of closed three-dimensional spaces. The most complicated (and first truly nontrivial) of these is called Bianchi IX. In it, the shape can change in characteristically different ways depending on the matter that it contains, if any. If there is none, the system is called vacuum Bianchi IX, which has two degrees of freedom like the Newtonian minimal model; if matter is present, the system will have three or more degrees of freedom. Just like the three-body problem, the Bianchi IX model has a shape sphere; the two are compared in Fig. 22. Please bear in mind that what you see there is not a snapshot at some instant of an evolving squashed three-sphere but a representation of its shape potential at each pair of the possible values that the Misner anisotropy parameters can take. To each pair there corresponds a point on the sphere; here too the northern and southern hemispheres correspond to mirror images. In both cases, the most symmetric shapes are represented by points at the poles: the equilateral triangle on the left and the so-called *round sphere* (the three-dimensional analogue of a soccer ball) on the right.

By analogy with the equilateral triangle, I shall call this latter shape Alpha. The equators correspond to degenerate situations in which only one independent distance ratio (and not two) can be formed. Just like the degenerate collinear triangles of the particle model, there are degenerate geometries. They come in two forms and are called *pancakes* when the geometry becomes two-dimensional and *cigars* in the doubly degenerate case analogous to two-particle coincidence.

Despite these similarities, there are differences whose importance means that they need to be highlighted. I will leave until later in the chapter the general situation, but it may be noted here that the Bianchi IX model has no analogues of the Euler configurations and there are just the three distinguished points of the cigars on the equator. Also, equal values of the corresponding shape potentials lie on contours that run in quite different ways over the respective shape spheres. The Newtonian contours tend to cut the equator orthogonally, while the Bianchi IX contours mostly run parallel to it except where they pinch into it at the 'cigars'. The only close parallel is the single distinguished point in each model; like the equilateral triangle, the round sphere is at the pole. But even here there is an important difference. There is no analogue of the unequal-mass shape sphere with equilateral triangle displaced from the pole. The biggest difference of all is the way the respective shape potentials vary over the spheres. The form of the potential

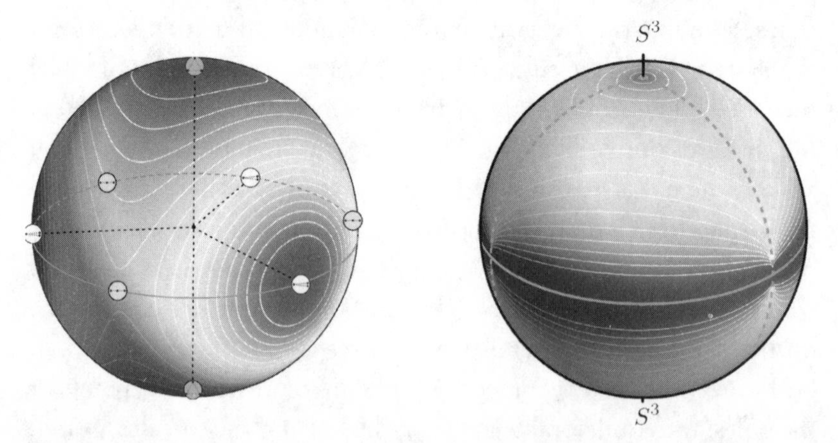

FIGURE 22. Equal-mass three-body (left) and Bianchi IX (right) shape spheres.

determines the dynamics; the difference between the potentials has profound consequences.

This is because we want to know where the size of the system (the volume of the universe, in Einstein's theory) can vanish. Quite contrary to the behaviour of the root-mean-square length ℓ_{rms} in the equal-mass particle model, it is quite impossible, except in purely homothetic motion in which nothing but the volume changes, for the system to reach the round-sphere pole S^3 on the right in Fig. 22 with zero volume through a motion in which the shape changes. It cannot reach that shape—Alpha, as I have called it—with zero size along a path in which the shape changes. Unlike Newton's equations, Einstein's do not permit nonhomothetic motion to total collapse of the size at the pole. They only allow shape-changing motion that drives the volume to a maximum size at Alpha. In such motion with increasing volume, the round sphere is an attractor like those in Chapter 11.

In a further major difference, the volume must (subject to a qualification to which I will shortly come) vanish at all points on the equator. In the particle model, that can only happen at the Euler configurations. Such configurations do not exist in Bianchi IX. Equally significant is that whereas the three-body trajectories can cross the equator without ℓ_{rms} vanishing, the volume must vanish if it reaches the Bianchi IX equator. This is the simplest nontrivial model of a big bang; unlike the Newtonian counterparts, it is generic. What is more, when discussing central configurations I mentioned that it is not known whether, for a given N, the number of them is always finite or if an infinite continuum of them exists. It is definitely known that none exist for $N = 3$ and $N = 4$ (the proof in this case is a notable feat) and also none for $N = 5$ apart possibly for some very limited cases. Beyond that nothing is known. In Bianchi IX, which corresponds to $N = 3$ in the particle model, the equator is analogous to a whole continuum of central configurations at which collapse to zero size is not merely possible but obligatory. With an eye to later discussion, I will call the equator Beta, reserving Alpha for the round sphere (not only in Bianchi IX but in all cases).

IN VIEW OF these differences, which are all essentially due to the different forms of the relevant shape potentials, one may wonder whether the *N*-body total collisions can offer any good guide to what happens in general relativity. I think they do, but before we come to that, some more preparatory explanation is needed. The critical question is not so much whether but how the solution curves in general relativity reach a point of zero size. In the more radical approach to the dynamics of the universe suggested by the *N*-body total-collision solutions, that is what we want a Janus point to be. It has long been known that the Bianchi IX solutions do invariably reach zero size at the equator, or at least they do when described in the standard spacetime formulation of general relativity. Let me first describe that. The behaviour of the solutions, without matter and also with most forms of it, is rather odd. As zero size is approached (going backwards in time to the big bang or forwards to a big crunch), the solution curves exhibit violent and chaotic changes of direction. The shape behaves like a ball bouncing repeatedly and randomly off a triangular pool table with infinitely many bounces before the size reaches zero. It does that in a finite amount of the time that Einstein's 'sinful' clocks, moving appropriately, would measure. However, in the regime in which this happens no useful clocks could exist. They would be pulled apart by the tidal forces exerted by the rapidly varying geometry. If one plots the trajectories of such Bianchi IX solutions as curves in their shape space, they go on forever. They never reach a Janus point.

This does not look good; a Janus point is the litmus test. Luckily the qualification 'with most forms of matter' opens up some wiggle room. I noted the impressive ability of Einstein's theory to accommodate many different forms of matter; here one comes to the rescue. It is called a massless scalar field. Modern theories of elementary particles suggest it might well be present at or near the big bang. Unlike electric and magnetic fields, which have both an intensity and a direction (which enables a compass to point to the magnetic north pole), a scalar field has only an intensity. Despite this simplicity, it has a decisive effect on the otherwise violent pool-table behaviour in Bianchi IX. Normally matter and geometry interact on equal terms, but in the vicinity of the big bang and in black holes there is something more like a wrestling match in which one contestant, the geometry, has the other, the matter, in a lock. Whatever the geometry does, the matter must follow suit. Wheeler, typically, summed it up with an apt pun: "Matter doesn't matter". To put it another way, matter is a helpless rider clinging to a bolting horse.

The exception that proves the rule is the massless scalar field. Actually, another is a 'stiff fluid'. Like an ordinary fluid that has measurable macroscopic properties but at a more fundamental level is composed of molecules in motion, a stiff fluid (in which the speed of sound is equal to the speed of light) is an emergent form of matter. Both it and a massless scalar field can take on geometry on equal terms. When either of these forms of matter is present, the otherwise violent and chaotic behaviour of Bianchi IX is radically changed. It is 'tamed' and said to become quiescent. This is because, tracing the evolution backwards in time to the big bang, there are no longer infinitely many bounces but only finitely many before zero size is reached. This property was known several decades before the paper of my collaborators mentioned earlier. However, the prevailing opinion was that the infinite densities that are still unavoidably reached heralded the breakdown of general relativity, and nobody seems to have considered the possibility that the variables which describe the shape of the universe might remain smooth and mathematically well behaved right up to the apparent singularity. They would be like one of the *N*-body back-to-back solutions that reach Alpha in a perfectly well-behaved way. Then, even if there is no unique continuation of the solutions through Beta, as I have called the Bianchi IX equator in Figs. 22 and 23, the apparent breakdown of general relativity would be a mere artifact reflecting unconscious transfer of concepts perfectly acceptable in the laboratory to the conditions at the big bang.

Proving this in all generality will be a major undertaking, about which I will say something soon, but my collaborators were able to make an promising start. Their aim, dictated by the way we conceived the nature of the Janus point at that time, was to make sure that, at least in Bianchi IX, a solution that exists right up to the singularity of zero size can be continued through it in a unique way as one single solution. In accordance with a well-defined mathematical criterion, they showed that this could be done. Thinking about the analogous situation in the *N*-body total collisions, in which that cannot be done, I have come to wonder whether the employed criterion is valid in the full big-bang situation. However, I don't think that need be significant. Back in 2015 we were not aware of the back-to-back total-collision solutions that, as conjoined twin bundles at an *N*-body Alpha, still respect time-reversal symmetry in a Janus-point fashion. What seems to me important now is the possibility of tracing the shape degrees of freedom all the way back to the big bang.

In his most recent book, *Fashion, Faith and Fantasy*, Roger Penrose points out that, in accordance with the current cosmological theory of inflation (which features in Chapter 18), it is not possible to 'see' what the big bang is like because the thermalisation which takes place as inflation comes to an end will "obliterate information about the actual nature of the [universe's] initial state". In fact, with or without inflation modern theory and observations enable cosmologists to determine the shape degrees of freedom of the universe just this side of the big bang. If we have a theory which enables us to trace them from there reliably back to the big bang, then we will know what it looks like. Perhaps a quotation from St Paul is not out of place here: "For now we see through a glass, darkly; but then face to face". 'Glass' in this King James rendering is a mirror, not the smoked glass one should use to look at a total eclipse of the sun, though that is actually more appropriate as a metaphor for trying to see the big bang. My conjecture, which is quite unlike Penrose's and to which I will come a little later, is that the big bang has a shape that everywhere looks somewhat like the parts of the 500-particle central configuration of Fig. 18 (page 212) not too near its spherical edge.

Fig. 23 is taken from my collaborators' paper but is slightly modified to show only the evolution up to the point of zero volume on the equator. It tells us not only that the Bianchi IX big bang is like a pancake, round if the collapse to zero size is at the midpoint between the cigars and more and more pointedly elliptical depending on the distance from the midpoint at which that happens. There is one important similarity with the Newtonian case that I want to emphasise especially. I said the scalar field makes Bianchi IX quiescent, so it undergoes only a finite number of bounces. The final bounce occurs at the last bend in the white curve before it becomes tangent to the black curve, which is a geodesic on the Bianchi shape sphere and corresponds to the simplest of the Bianchi models (Bianchi I). Here too we see that as the size in the extended representation tends to zero, the behaviour of the system in the reduced representation in shape space becomes simpler. Just as in the Newtonian model, this suggests to me that quantisation of general relativity will be easiest in the vicinity of the big bang. The conditions there look benign.

Although Einstein's and Newton's theories match up in this respect, they do not when it comes to the angle with which the white curve leaves the equator. We have seen in Chapter 16 how the N-body problem nonhomothetic solutions are, in the absence of contingent symmetry, forced to leave the Janus point along one of the eigendirections of the Hessian of the shape potential (the great bulk doing so along the one with largest eigenvalue). We do not see the same behaviour directly visible in the topography of shape

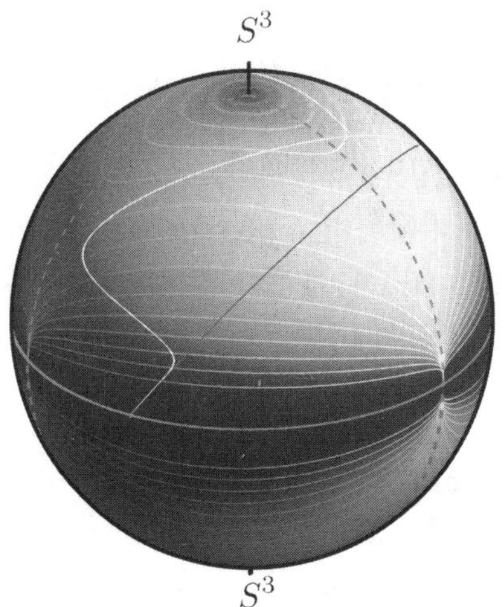

FIGURE 23. Bianchi IX (white) and Bianchi I (black) solutions.

space because we are not dealing with pure geometry, as would be the case if there were no scalar field, such as in vacuum Bianchi IX. When the massless scalar field is added, it augments the two degrees of freedom that vacuum Bianchi IX possesses. It plays a role in the dynamics, but because it is homogeneous the contours of the Bianchi IX shape potential do not in any way reflect its presence. That is manifested in the behaviour of the white curve in Fig. 23; together with the geometry, it determines the angle with which the curve leaves the equator. After that an analogue of the creation measure comes into play, and the triumvirate it forms with the geometry and scalar field determines how the curve then bends. It would be good if we could depict the scalar field in topography, but we cannot as long as it remains homogeneous. That renders it as invisible as Newton's absolute space; with the geometry, it cannot codetermine contours in a new arena that geometry and matter together create.

BEFORE WE CONSIDER what that could be like, we must grapple with generic solutions; those of quiescent Bianchi IX are too simple to serve as adequate guides. We confront a daunting task but not, I think, one that is completely hopeless. In the *N*-body problem there are reasonably good grounds to believe we know what happens when the number of particles is increased. There will be a great increase in the number of central configurations as *N* is increased step by step, but provided there is no continuum of central configurations their number will remain finite. In contrast, the number is already infinite in quiescent Bianchi IX, in which every point on Beta, the equator, is a Janus point; that is in a dynamical system with only one degree of freedom more than the particle minimal model has. But quiescent Bianchi IX already represents a reduction in degrees of freedom from infinity to three. Won't the behaviour at the full Einsteinian big bang be infinitely more complicated than in quiescent Bianchi IX?

The structure will surely be vastly richer in fine detail—the information-theory number of bits needed to communicate the structure of the 500-particle central configuration of Fig. 24 greatly exceeds the two Jonah will need to tell me he holds the equilateral triangle in his hand. There is, however, no difference in kind. Both structures maximise uniformity to the extent that the attendant circumstances—the mass ratios in Newtonian theory—allow. There is one simple reason why *N*-body total collisions with their ineluctable destinations at central configurations—and total explosions leaving them—may be a guide to the big bang, and it is that in both cases the size of the universe goes to zero. We have seen the royal effect that zero size has. An image that comes to mind is a champion sheepdog on the open fells chivvying recalcitrant sheep into a pen in the valley. Is there an analogous effect in general relativity?

The first evidence that there might be appeared in Russia in 1970 in work by Vladimir Belinskii, Isaak Khalatnikov, and Evgeny Lifshitz; their main conclusion came to be known as the BKL conjecture. It was the work of BKL, as I shall call them, which led to recognition of the 'matter doesn't matter' effect near certain kinds of singularities. The exception to this rule that makes quiescence possible was soon recognised, though not given great weight.

More significant for the present discussion is *asymptotic silence*. To explain what that is, I need first to recall from Chapter 8 the definition of simultaneity in general relativity. As I illustrated with wavy slices through a loaf of bread, there is no unique notion of simultaneity, only one defined by a spacelike hypersurface that can be chosen with a great amount of freedom. In the loaf analogy, with speed of light set equal to unity, the only restriction on the slices is that their slope must be nowhere more than 45°. In a spatially closed spacetime, which on Machian grounds I assume throughout, the big bang will, provided reasonable conditions are satisfied, take place simultaneously on a spacelike hypersurface. If the BKL conjecture is correct, and there is increasing evidence which indicates that it is, there are striking implications for the structure of spacetime. As timelike geodesics (ones inclined from the horizontal in the 'loaf' by more than 45°, i.e., either toward the past or toward the future) pass through spacetime, they behave in a very special way if traced toward the spacelike hypersurface on which the big bang is located. Away from it the path a geodesic takes is determined by the structure of the spacetime around it not only forwards and backwards in time but also in the three directions of space. Neighbouring geodesics 'feel' their shared spatial environment and in a sense are in communication through it. However, as the big bang is approached the BKL conjecture suggests this communication dies out and neighbours can no longer 'talk' to each other. This is what is meant by asymptotic silence. As the volume of the universe tends to zero, the dynamical evolution at each space point becomes independent of what is happening at neighbouring points. The upshot is that, with and without quiescence, Bianchi IX behaviour unfolds independently at each space point. If there is quiescence, and a research paper discussed in the notes provides evidence that it does, then in a goodly number (what that means is explained in the notes) of solutions behaviour like that depicted in Fig. 23 occurs at each point in the immediate vicinity of the big-bang spacelike hypersurface.

This behaviour is very like what happens in a nonhomothetic total collision. There the forces between individual particles seem to become nonexistent but in reality 'conspire' to force the shape of the system to that of a central configuration at the same time as the root-mean-square length ℓ_{rms}, the counterpart in particle dynamics of the volume in general relativity, becomes zero. This immediately prompts the question of whether the shape of the spacelike hypersurface on which the big bang occurs has some special structure like that of central configurations. This is plausible—what the royal zero does may be similar in the two cases. If it is, I think the big-bang hypersurface will have a structure in which both the scalar field and the two independent spatial curvatures vary from point to point, not violently but in a relatively uniform manner. The *N*-body central configuration of Fig. 18 (page 212) suggests the effect will be more like the distribution on the right of Fig. 24 (already seen as Fig. 11) than the distribution on its left. The dots in the figure are to be understood to represent locations at which the strength of the scalar field and the spatial curvatures are greater than at neighbouring points. It is to be expected that other fields will be present, but the 'matter doesn't matter' syndrome means their dynamical effect in the vicinity of the big bang can be ignored.

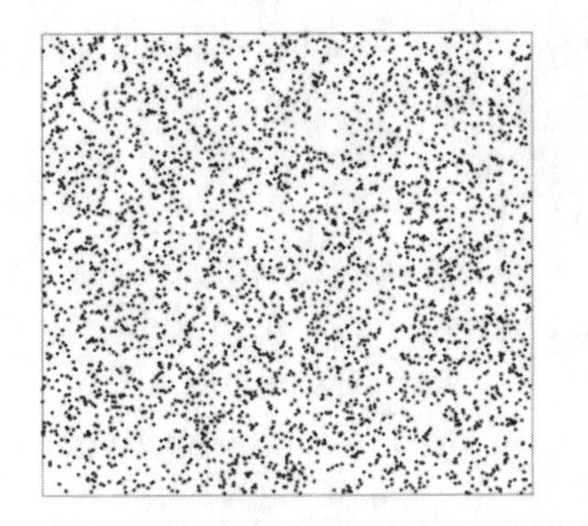
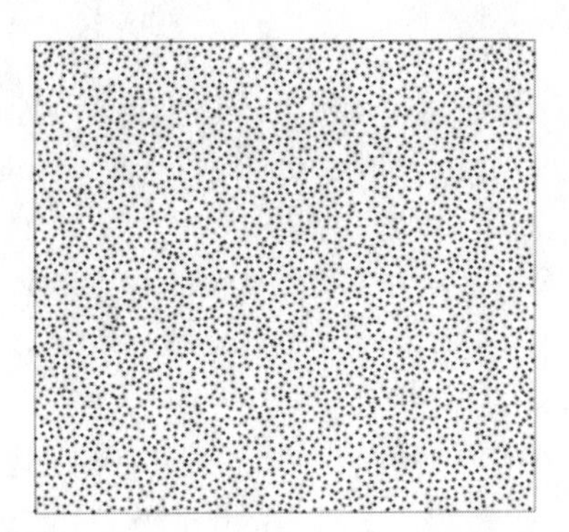

FIGURE 24. Poisson (left) and glassy (right) distributions of point particles.

If the picture just presented is in essence correct, there may be implications not only for the problem of time's arrows but also for research in general relativity. The existence of infinite densities at the big bang and in black holes follows from famous 'singularity theorems' proved in the 1960s and 1970s by Roger Penrose and Stephen Hawking. There is clearly no error in the mathematics that underlies their theorems. But if, as I have just suggested and will be considered further in Chapter 18, the shape (as opposed to the volume) of the big-bang spacelike hypersurface has some well-defined structure analogous to total-collision central configurations, the singularity theorems may have been given the wrong physical interpretation. The big bang will be not the demise of general relativity but its structural pinnacle.

Will it resemble Alpha of *N*-body theory or Beta of quiescent Bianchi IX? I have no idea, but I believe mathematical research can find the answer. In this chapter I have been a bit cavalier in introducing, without further ado, the shape potential and its contours as laid out over the geometry of Bianchi IX. It is actually the simplest realisation of a significant triumph of pure mathematics in the twentieth century, for which some background is needed. When I described in Chapter 8 how general relativity can be derived in a Machian manner, I explained that it is not particle configurations which change dynamically in Einstein's theory but curved three-dimensional spaces that close up on themselves as the surface of the earth does in two dimensions. They are called Riemannian three-geometries because Riemann introduced them along with counterparts in any number of dimensions in the lecture in 1854 that impressed Gauss. The amount of curvature at any point in such a three-geometry is denoted by R^3 and called the Ricci scalar (scalar because it is a single number); the Italian mathematician Gregorio Ricci-Curbastro found the convenient expression for it in what is called tensor analysis.* In the dynamics of geometry, at each space point the Ricci scalar plays the role of the Newton potential V_{New} for particles distributed anywhere in Euclidean space. That the value of R^3 is associated with a point (or, more precisely, the infinitesimal neighbourhood of a point, since spatial derivatives are needed in the definition of R^3) reflects the way Riemann retained Euclid's geometry 'in the small' and made a patchwork of infinitesimal Euclidean regions into the fabric of space. In the dynamics of space, potential energy is now associated with each point of space, not with bodies distributed in space.

AFTER THIS PREAMBLE I can now explain what the success in pure mathematics in the last century was. In 1960, the Japanese mathematician Hidehiko Yamabe claimed he had proved that, given any Riemannian geometry with dimension three or greater, it would be possible to convert it to a related Riemannian geometry in which the Ricci scalar has the same value in the entire space. This would be achieved by a conformal transformation of the kind that plays a crucial role in the Machian derivation of general relativity. You may recall it is one that leaves angles between intersecting curves unchanged but can increase or decrease the infinitesimal amount of space everywhere; for a two-dimensional illustration, think of the surface of one of those balloons that expand by different amounts in different places. If the angles with which curves on the surface intersect are not changed as you puff into the balloon, you have brought about a conformal transformation. Eight years after

---

*If you search online for 'Ricci Italian mathematician', as I did, all the first results refer to Matteo Ricci, the famous Jesuit missionary to China of the late sixteenth and early seventeenth centuries.

Yamabe had presented what he claimed was a proof of his assertion, it was found to contain a critical error. Luckily for Yamabe the combined efforts of three mathematicians had finally vindicated his claim by 1984.

If I and my collaborators are right in our belief that it is shapes that define the world, a pure number associated with what was originally Yamabe's conjecture is of the highest importance. Before saying why, I must first tell you what the number is. Curvature is characterised by a radius—the curvature of a circle at any point on its circumference is the distance from that point to the centre of the circle. Curvature has the dimensions $[l]^{-1}$ of an inverse length and the Ricci scalar those of an inverse length squared: $[l]^{-2}$. At the same time, the volume V of a three-geometry has the dimensions of a length cubed: $[l]^3$. Therefore if the Ricci scalar R made constant by the Yamabe transformation is multiplied by $V^{2/3}$, the two-thirds power of the volume the three-geometry has after the transformation, the mathematical alchemy of the twentieth century turns base metal into gold—it turns dimensionful curvature into a dimensionless pure number known as the Yamabe invariant. Just as the complexity, and with it the shape potential, is an invariant of the similarity group of Euclidean geometry, the Yamabe invariant plays the same role in Riemannian geometry. Sophistication is the word for it. It is an invariant of what I call the *geometrical group* and define in the notes. As regards the numbers of numbers that are in play, I think Newton would be pleased—it's *ex finito ad infinitum*.

It is not, however, what I think is needed if we are going to get an idea of what the big bang really does look like. The point is that the Yamabe invariant, although it does give much important information about what solutions are possible in general relativity, is a construct of pure geometry. There are also issues related to the fact, mentioned in the notes, that three classes of three-geometries are associated with it. The main thing of interest at this stage of the discussion is that in one class the Yamabe invariant has its maximum value, which is positive, at the round sphere and can have all values from that maximum down to minus infinity. It is immediately tempting to liken these decreasing values of the Yamabe invariant to those of the shape potential and its 'partner', the complexity. It is certainly the case that three-geometries with values of the Yamabe invariant less than the maximum will be more structured than the perfectly smooth round sphere.

But more is needed: matter. Somehow or other I think the notion of the Yamabe invariant must be extended to include matter and obtain a combined quantity (let me call it *Yamabe+Matter*) that is an invariant of the *universal group*, also defined in the notes. At the very least it would be helpful to include a scalar field, the one form of matter that can take on geometry on equal terms—though now no longer merely homogeneous, as in quiescent Bianchi IX, but with values that can vary from point to point. Then at each space point there will be two kinds of potential energy: the purely gravitational Ricci scalar and the potential energy of the scalar field.

If this combination of at least scalar matter with geometry—or, better, an all-encompassing Yamabe+Matter—has a unique extremum surrounded by contours that define directions analogous to the eigendirections in the *N*-body problem, I think we will know what the big bang looks like, not darkly but face-to-face. If, despite being agnostic, I may continue to use religious language, we would see the *locus iste*. These are the opening words, the incipit, of the Latin gradual used in Germany for the anniversary or dedication of a church; the full phrase is *Locus iste a Deo factus est*, which translates as "This place was made by God".

CHAPTER 18

THE LARGE-SCALE STRUCTURE OF THE UNIVERSE

THE PREVIOUS CHAPTERS HAVE SHOWN THAT THE JANUS-POINT MODEL DOES plausibly predict an initially uniform state of the universe that then evolves into one in which matter is clumped. That's good—it's what we see in the universe and know about its history—but it's only a start. Observations made in the last three decades have shown that the observed clumping has a rather definite nature and has arisen from an early state that was very uniform but with a distinctive and specific kind of order. At the current time, this order is attributed to what is known as *inflation* and an associated quantum mechanical effect that is believed to have generated inhomogeneities from which the large-scale structures in the universe have arisen. There are three observations that inflation explains rather impressively, one especially so, though by no means from clear first principles. I will first describe what the observations are and how the current theory of inflation explains them. I will then put forth the tentative alternative to inflation suggested by the discussion in Chapters 16 and 17.

The three things that inflation explains are best described in terms of the simplest cosmological models with a big bang first described in the 1920s and early 1930s. In their simplest form, involving only dust particles, they are the counterparts of the Newtonian homothetic solutions from which I said my grandson, imagined unable to see scale and empty space, could only identify an unchanging equilateral triangle. With a more sophisticated form of matter that includes radiation and is referred to as the cosmological fluid, they are the foundation of cosmology and are called FLRW solutions after the various researchers who, in chronological order, contributed to their development: the Russian meteorologist Alexander Friedmann, the Belgian Catholic priest George Lemaître, Howard P. Robertson, and Arthur Walker. The FLRW solutions describe the homogeneous and isotropic evolution of a cosmological fluid with different possible relationships between its energy density and pressure. They exist in three distinct forms depending on the structure of space (not spacetime) in them: it can be closed in three dimensions, like the surface of the earth or a soccer ball in two dimensions; flat, like a sheet of paper that extends to infinity (other possibilities exist, like a sheet rolled into a tube, but we do not need to consider them); or analogous to a saddle in a mountain pass that, unlike a horse's saddle, extends to infinity. In the early stage of evolution near the big bang, the fluid in the FLRW solutions describes the behaviour of radiation, which is the dominant dynamical component. However, the expansion of the universe causes the radiation energy density to decrease as the inverse fourth power of the time, whereas the energy density of the matter decreases only as the third power. At a certain epoch it therefore takes over from the radiation in controlling the dynamics. The FLRW solutions, made more realistic by the introduction of small inhomogeneities, are the workhorses of cosmology.* Already in the 1970s they brought to the fore three major issues.

---

*There are, I believe, two reasons, to have reservations about the FLRW solutions. First, they may give a distorted picture due to failure to correct for Einstein's 'sin' (page 247) of not deriving clocks and rods intrinsically from the equations of the theory, and therefore not having a true theory of time and scale in the foundations of his theory. Two things, an independent variable and a dependent variable, are the minimal prerequisite of any dynamical theory. Time, taken as given, is generally the former, while whatever is being considered, say the height above the ground of a falling apple, will be the latter. In the FLRW solutions, the independent variable is called proper time, but no real clock measures it. There are three dependent variables: the

The first is the *horizon problem*. After Hubble's discovery that the universe is expanding, the next major discovery in cosmology came by accident in 1964 when it was found that the universe is bathed in thermal radiation at the very low temperature of 2.73 kelvin. It has approximately the same wavelength as is used in microwave ovens and is called the cosmic microwave background (CMB). It was immediately recognised to be an 'echo' of the big bang and must have had a much higher temperature in the past that has since been steadily reduced by the expansion of the universe. The problem arose because observations showed that the temperature of the microwave background radiation is extraordinarily uniform, with fluctuations of only about one part in 100 000 across the sky. Such a striking effect called for an explanation, and it was rather natural to assume it had been established causally through equilibration. Ever since the work of Boltzmann, that was how thermal equilibrium was understood to come about. However, this assumption led to a conflict between the homogeneous and isotropic FLRW cosmological models and the fact that in accordance with relativity theory no physical effects can be propagated faster than the speed of light. It was realised that, when traced back in accordance with the FLRW models, no two points in the CMB separated on the sky by more than 1° (twice the angular diameter of the moon) could ever have been close enough to each other for light and causal effects to have connected them. They would have been outside their horizon of possible causal interaction.

---

scale factor, the energy density, and the pressure of the cosmic fluid. For these too there is nothing to measure them properly. Cosmology is nevertheless an impressively successful science because the predictions it makes with these potentially suspect tools can be checked rather well by numerous concordant observations. However, these do not protect the theory against possible errors of two kinds: first, near the big bang (where no clocks and rods can exist), and second, a very slow drifting apart of the assumed increase of cosmological time and of terrestrial atomic clocks. It is not impossible that the apparent accelerated expansion of the universe is an artifact of such a drift. The second reason to have doubts about the FLRW solutions is that they may miss the most important distinction between possible cosmological solutions that arises from the fact, illustrated in Fig. 21, especially on the right, and the discussion accompanying it, that different values of the creation measure give rise to different amounts of structure formation in the universe. There is nothing in the formal structure of the FLRW solutions that reflects this role. It is not easy to calculate the effect of structure formation on the expansion rate.

The problem can be illustrated by ants, taken to represent light, that can move at a certain speed over the surface of an expanding balloon. Two marks on the surface can move apart so fast that an ant can never get from one to the other. At a given stage in the expansion there will for any mark (representing an observer's position) be a horizon of points on the balloon beyond which no ant could have reached the mark in the time since expansion of the balloon began.

The second issue was recognised because observations of the matter in the universe indicated that the energy density in the universe was, if one assumed it to be determined by pure chance, surprisingly close to the critical value for the case of a flat universe. This was the *flatness problem*. It is best explained in terms of the famous physical units that Planck introduced in 1899 just before his discovery of the first quantum effects. By then he was already aware that a hitherto unrecognised constant, with the dimensions of angular momentum and now called Planck's constant, must play a role in the explanation of the so-called blackbody radiation. Using combinations of this constant, the speed of light, and Newton's gravitational constant, Planck was able to define what are now called the Planck length, time, and mass. They are of great interest for two reasons. First, since they are associated with universal properties of nature, intelligent beings anywhere in the universe could use them to communicate the results of physical measurements to other such beings wherever they might be. Second, by terrestrial standards both the Planck length and Planck time are incredibly small, 5.39×10^{-44} sec and 1.62×10^{-35} m, respectively. In contrast, the Planck mass (or energy, by the equivalence Einstein established) is very large compared with the masses of the well-known elementary particles such as the proton.

The widely accepted significance of the Planck units is that currently known physics will cease to be valid on scales of their magnitude. The flatness problem, like the problem of the incredibly small value of the cosmological constant, can be formulated in various ways using the units. One is in terms of the total energy budget of the universe at the Planck scale, using a flat universe with matter fields on it representing both curvature and the cosmological constant. At this scale the fraction of energy that is in the cosmological constant is one part in 10^{122} (much evidence indicates it is not exactly zero), and the fraction that is in the curvature of space is not more than one part in 10^{60}. These extremely tiny fractions are extremely hard to understand. Not unrelated to this consideration is the colossal size, compared with the Planck length, of the universe in its current epoch as measured by the Hubble radius (the distance from an observer at which the recession velocity of a galaxy equals the speed of light).

Before continuing with the third issue that led to the theory of inflation, I will take the opportunity to say something about the Planck units. They reflect what are probably the three most fundamental facts about the physical phenomena we observe around us: all bodies accelerate at the same rate in a gravitational field, nothing can travel faster than light, and the Planck constant brings discreteness into the world; it is the origin of distinct things we can count and with that pure numbers in physics. Neither gravity nor light gives us that valuable gift; instead, each in its own way reflects the profound universality of behaviour. Planck's constant is the odd one out, sui generis. It is currently taken as a given that we identify in local phenomena.

You will have noticed that I have taken pains in this book to show, in several examples, how local phenomena emerge from a law of the universe at epochs sufficiently far from the Janus point, which we can call the big bang if the universe has zero size at it. Is it possible that Planck's constant is emergent?

Here a possible clue is that it has the dimensions of angular momentum. The greatest step toward fully fledged quantum mechanics surely came when Niels Bohr suggested that electrons could only circle the nuclei of atoms in orbits that are fixed multiples of Planck's constant. Though this is impossible to understand from a classical point of view, it gave an immediate explanation of the discrete energies atoms emit as radiation. The first success was for the atoms of hydrogen, then helium, and ultimately on to all quantised effects in physics. Also noteworthy in this connection is what

Bekenstein, as we saw in the discussion of black hole entropy in the final part of Chapter 15, was forced to do in order to convert the dimensions of area that a black hole's event horizon has to the correct dimensions of entropy when it is measured in energy units. He had to use a quantity with the dimensions of angular momentum; the only available quantity was Planck's constant. We have seen the sensation to which that, through Hawking's intervention, soon led.

If, in the light of these comments, we are prompted to look for quantities with the dimensions of angular momentum in a relational universe, we find them in precisely two places. There is none in overall rotation because the angular momentum of a relational universe is and forever remains exactly zero. But the universe does—indeed, must—have the dilatational momentum D, the measure of the amount of kinetic energy it has in overall expansion. In fact, the ratio of the kinetic energy in change of shape to that in overall expansion is what I have called the creation measure, which I have surmised is the most important number in the universe. Thus, D is one quantity with the dimensions of angular momentum. Is it possible that there is a deep, hitherto unsuspected connection between Planck's constant and the quantity analogous to D in general relativity?

The third issue that exercised cosmologists in the 1970s concerned the origin of galaxies and clusters of galaxies. All cosmologists knew that they must have formed by gravitational clumping but that this would require a much more sophisticated theory than a simple discussion based on the N-body problem. The main shortcoming of a Newtonian particle model in the early universe, up to about 300 000 years after the big bang, is its complete neglect of radiation and the associated effects related to the speed of light. As I said earlier, it was radiation, not matter, that determined the expansion rate of the universe at early times. Moreover, clumping of matter depended on competition between gravitational attraction and repulsion due to the pressure of the fluid-like medium that filled space. Regions of higher density (overdensities) could grow only if they were pronounced enough for gravity to overcome pressure. In the framework of the completely isotropic and homogeneous FLRW models it was not even possible to study the growth of overdensities since some must exist as an initial condition before one could begin to calculate how they would evolve. There had to be fluctuations, deviations from uniformity.

In the early 1970s, in the absence of anything better, a British cosmologist working in the United States, Edward Harrison, and the Russian physicist, Yakov Zel'dovich, independently proposed what seemed to be a simple and plausible form the fluctuations might have had at some stage in the very early universe. Ever since the great work by the French mathematician Joseph Fourier it has been standard practice in physics to represent departures from uniformity by superpositions of perfectly sinusoidal waves of all possible wavelengths. Harrison and Zel'dovich proposed for the fluctuations what is called a scale-invariant power spectrum. Scale invariance here is not to be confused with the sense in which the complexity is independent of scale but refers to a specific relationship that holds when the wavelength and amplitude of the Fourier waves are plotted against each other in the limit when the wavelengths tend to infinity. This gives what is called the power spectrum. A straight line with a 45° slope is obtained.

It will not be easy for a reader unversed in the art of Fourier representation, which is extremely useful when one needs to calculate the initial evolution of fluctuations, to get an idea of what a Harrison-Zel'dovich spectrum corresponds to as a distribution in real space. You can get a rough sense from the distributions I showed in Fig. 11 (repeated as Fig. 24, page 259) to illustrate the ability of the complexity to distinguish distributions of particles that are clustered to a greater or lesser extent. The distribution on the left of Fig. 24 is the paradigmatic example of what can be observed when independent random events occur and is a Poisson distribution; it is generated point by point, with each successive dot appearing with equal probability anywhere on the square irrespective of where previous dots have appeared. There are more voids and coincidences than one would intuitively expect. In the distribution on the right of Fig. 24, called glassy because it is typical of the distribution of molecules in glass (they are not at the points of a regular lattice), you get an idea of what Harrison-Zel'dovich fluctuations in the cosmological fluid look like. The points represent the positions of the overdensities in the fluid and they exhibit what is called long-range order. This means that the appearance of the fluctuations in one localised region chosen at random looks much more like any other region chosen at random far away than is the case for the Poisson distribution. That is very evident in the figure.

Given a Harrison-Zel'dovich spectrum, the decisive thing for the growth of structure in the universe is the relationship between such a fluctuation spectrum and the horizon of possible causal interaction discussed earlier. During expansion of the universe from a big bang, the diameter of causally connected regions will have a certain value. Some of the fluctuation wavelengths will be greater or less than it. Causal processes will affect the fluctuations with wavelengths within the causal horizon. Given preexisting Harrison-Zel'dovich fluctuations, FLRW expansion of the universe, and certain conditions of the matter, it was possible by the 1970s, above all through work by Joseph Silk and James Peebles, a recent Nobel laureate in physics, to predict detailed properties of the CMB decades before they could actually be measured. In turn, these properties are reflected in the distribution of galaxies and clusters of galaxies within the universe. Indeed, on a large enough scale the observed distribution bears the imprint of the Harrison-Zel'dovich spectrum, though with a slope slightly less than 45°.

THERE IS, OF course, also the question of the mechanism by which the Harrison-Zel'dovich fluctuations might have arisen in the first place. The inflationary proposal, now widely accepted by cosmologists, arose in a remarkable indirect manner. As is often the case in major scientific discoveries, numerous people made important contributions, but the initial stimulus undoubtedly came in the late 1970s from the young American Alan Guth. His doctorate had been in high-energy particle physics, in which dramatic advances were being made at that time. Exciting new theories, as yet without experimental support, had been proposed. Some suggested that there should exist hitherto never observed magnetic monopoles. This was a radical departure. Isolated positive and negative electric charges had long been known to exist, but magnets always have a north pole and a south pole. If one attempts to break a magnet in two in order to separate the poles, the resulting halves each have a north pole and a south pole.

Guth was concerned by the fact that if the proposed theories were correct, magnetic monopoles should have been created in great numbers under the high-temperature conditions very soon after the big bang. Why were none to be found today? Where had they all gone? Guth realised that the very theories which predicted the existence of the magnetic monopoles would also give rise to a form of matter, a so-called false vacuum, that would have an anti-gravitational effect and cause the expansion of the universe to be accelerated rather than decelerated by the usual effect of matter. If accelerated expansion lasted long enough, this could solve the problem of the missing monopoles: they would be spread out so widely through the universe that we could not reasonably expect to observe any now.

Guth was also aware of the horizon and flatness problems and was naturally excited to realise that the accelerated expansion could solve both of them as well. First, the expansion would have the effect of increasing the size of the universe so much that, in my analogy of ants on the surface of a balloon, the universe would to all intents and purposes be flat. His solution to the horizon problem was as follows. Suppose that very soon after the big bang there emerged a region small enough that within it, whatever the initial conditions, there had been time for thermal equilibrium to be established. If there was now to be extremely rapid accelerated expansion for long enough, the region of uniform temperature would be spread over a region large enough to explain why we observe almost exactly the same temperature all over the sky.

Guth soon started to get positive responses from scientists with whom he discussed the idea. He was, however, worried about the manner in which the accelerated expansion could come to an end. It threatened to spoil his beautiful theory. He called it "the graceful exit problem". Despite this worry, he decided to publish a paper. It appeared in 1981 with the title 'Inflationary universe: A possible solution to the horizon and flatness problems'. The term Guth felicitously coined, 'inflation', has stuck. At the time of writing, the paper has had over 11 000 citations. The absence of 'monopole' in the title suggests a certain amount of what might be called mission drift. In fact, much more drift was to come but, far from being a case of a campaign that got bogged down, this turned out to be a triumph. There's a blow-by-blow account of the whole story in Guth's book *The Inflationary Universe*, published in 1999.

Two decisive steps led to the triumph within a year or two. The first was replacement of the form of matter responsible for inflation—the accelerated expansion of the universe. Whereas matter fields—for example, the electromagnetic field—generally have the familiar attractive effect of gravity, it turns out that the simplest of all possible fields, a scalar field of the kind that makes the Bianchi IX model of Chapter 17 quiescent, can have a repulsive effect, provided certain conditions are satisfied. In general, such a field will, like all forms of matter, have at any instant kinetic and potential energy. In the present forms of the theory it is usually assumed that the latter is dominant, very large at the onset of inflation and, in what is called slow roll, decreasing steadily as it drives accelerated expansion of the universe. The final stage of the process, when the potential energy is largely exhausted, can take place gracefully, thereby solving the nagging problem that Guth still had when he published his paper in 1981. In fact, in a further example of mission drift, the state of thermal equilibrium observed in the CMB is now assumed to be established not before the onset of inflation but at the end. In an example of linguistic inertia that reflects the historical development, this is called 'reheating', although the original preinflation 'heating' (through equilibration in a very small region) no longer features in the overall picture (or at least not in the accounts of the present state of inflationary theory I have read).

Although the theory of inflation has undoubted strengths—I am just about to get to the one that underlies its widespread popularity—one of its weaknesses is that the precise nature and properties of the field that drives inflation cannot, at the moment, be deduced from any kind of sound theory. In fact, one hypothesises a field, now called the inflaton, that has the properties needed to explain the observed facts. It would be quite wrong to say that inflationary theory is ad hoc, but parts of it are post hoc. For example, the properties of the inflaton have been fixed to ensure that enough inflation does occur to resolve the flatness and horizon problems; it is, however, good that enough inflation to solve one problem will solve the other. But this positive feature is far less significant than the great triumph of inflation and the reason for its central role in modern cosmological theory. This is because it provides a mechanism that, very neatly, can generate the all-important Harrison-Zel'dovich fluctuations, which in turn can explain the structure of the universe on its very largest scales. The mechanism relies on the interplay between quantum and classical physics. I think the most effective way to explain this is to sketch the picture now generally presented in broad-brush accounts of inflation.

AS NOW PRESENTED, it is assumed that 'at the beginning' the yet-to-be-created theory of quantum gravity, the unification of Einstein's general relativity with quantum theory, holds sway. Out of some quantum state variously described as 'spacetime foam' or 'quantum soup', a spacetime that is effectively classical emerges within a few Planck times of a presumed birth of the universe.* Within this spacetime at least a part begins to inflate exponentially, doubling in size repeatedly and very rapidly. Because of this behaviour, the resulting spacetime approximates closely a solution the Dutch astronomer Willem de Sitter found in 1917 for the field equations of general relativity in the form modified with the so-called cosmological constant that Einstein had introduced in 1916 in a flawed attempt to implement Mach's principle. As in Guth's original

---

*Besides a Janus-point origin, there are now several proposals—involving bouncing universes or ones that expand after a long quasi-stable period—in which the universe does not explode into existence. These have often been created with properties tailored to be rivals for inflation.

proposal, this has the potential to solve both the horizon problem and the flatness problem.

Critically, this is not the full story. It is reasonable to assume that quantum mechanical fluctuations of the inflaton are superimposed on the classical background. In the simplest solution of Einstein's equations, flat Minkowski space, there exists the well-tested quantum field theory, the key element of which is its lowest energy state, known as the vacuum state. Various vacuum states can exist in de Sitter space; one of them, called the Bunch-Davies vacuum, is singled out by a rather natural condition. It consists of fluctuations of all different possible wavelengths combined in a way to create a quantum state that overall has the same symmetry as de Sitter space. As the background spacetime inflates, the vacuum state is deformed in a very specific way that leaves its symmetry unchanged but radically changes the structure of the fluctuations of which it is composed. They undergo the quantum mechanical process known as squeezing. This is all very striking because the resulting fluctuations, magnified from their original microscopic quantum mechanical size up to a macroscopic scale, have exactly the form of the Harrison-Zel'dovich fluctuations. Moreover, in the simplest models they typically have a slope slightly less than 45°, as is actually found in the analysis of the CMB observations.

With good reason, most cosmologists think this is a wonderful result—quantum effects at the smallest scales can, together with other parts of physics, explain the form of the very largest structures in the universe, which are galaxies and clusters of galaxies. Guth had nothing like this in mind when he started to worry about magnetic monopoles or even when he published his paper. It is particularly impressive that his direct attempt to solve three major issues—the monopole, horizon, and flatness problems—should, as a remarkable by-product, also explain the origin of the large-scale structure of the universe.

Nevertheless, some people do have doubts about inflation. First, there is the post hoc aspect I already mentioned. There are endless debates about the possible properties of the inflaton. As yet there is little in the way of a first-principles derivation of them.

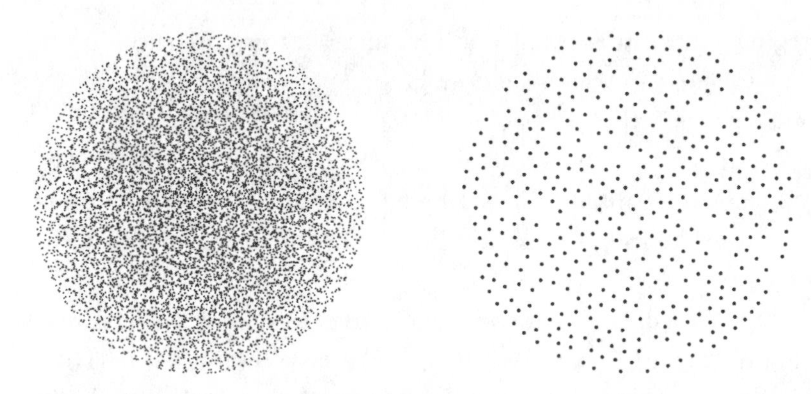

FIGURE 25. The central configuration shown earlier has a very uniform distribution.

To quite an extent they are chosen to give the observed properties of the CMB. There is, however, one model in which this is not the case. A couple of years before Guth published his paper, the Russian Alexei Starobinsky considered the simplest possible quantum-inspired modification of Einstein's equations and obtained a quite definite spacetime with an effective slow-roll inflaton that fits the CMB observations rather well. At the time, Starobinsky had not, unlike Guth, thought to resolve the flatness and horizon problems by the exponential expansion his model predicts.

Despite this positive aspect of Starobinsky's model, there is still no definitive theory of quantum gravity to support it or any other proposals. Even more problematic is the question of how inflation gets started. This depends critically on the nature of the spacetime assumed to emerge out of the primordial quantum soup. There is widespread agreement that inflation cannot begin without there being sufficient smoothness in the universe at the time when it can first be regarded as effectively classical. Current research aims to establish the necessary degree of smoothness and its extent. Generally people seem to think that the classical universe is likely to be very non-uniform when it comes into existence. If so, inflation might never happen. One answer to this problem is called chaotic inflation. The idea is this: if the universe has infinite extent, then even if its state is very non-uniform there must somewhere be regions uniform enough to allow inflation to occur within them. The universe in which we now find ourselves will then be part of such a region that was able to inflate. This kind of reasoning involves issues around probabilities when infinitely many possibilities can be realised; this is where the issue of Boltzmann brains comes to the fore. I think it is fair to say the related problems are as yet unresolved.

What I will anticipate here and return to soon is the possibility that the law of the universe makes it almost certain that it will be born (at the Janus point) in a very uniform state. If the Newtonian total collisions serve as any useful guide to the big bang in general relativity, that is what they do suggest. To emphasise that point, I repeat here in Fig. 25 the figure, shown earlier as Fig. 18, that depicts the distribution of the particles in a typical central configuration near the maximum of the shape potential. The distribution is very uniform. At least to the eye it looks not unlike the glassy distribution on the right of Fig. 24. Of course, the distinguished centre of mass means this *N*-body model cannot describe the real universe, but as I argued when introducing the model, the region near its centre can, if there are sufficiently many particles, model the conditions in a typical region of the universe. In fact, for the reason explained in the notes to Chapter 16, the distribution hardly changes all the way to the rim.

Finally, some people also worry how the quantum mechanical Bunch-Davies vacuum, squeezed by the inflationary expansion, becomes classical. This, like the pre-inflationary transition from a quantum soup to an effectively classical spacetime, is related to the notorious measurement problem in quantum mechanics made famous by Schrödinger's unfortunate cat, which is simultaneously both alive and dead until somebody opens its box and 'collapses its wave function' into one of the two possibilities. Modern-day experimental physicists have not the remotest chance of performing such experiments, even though theory and all experiments hitherto made show that it should in principle be possible. An experiment in 1999 with carbon-60 molecules ('bucky balls') arranged to be simultaneously in two different positions in space at once confirmed the quantum-theoretical prediction, but that was still only for relatively large microscopic objects. That's a long way short of mesoscopic grains of sand, let alone cats. One possibility is that, as suggested in the many-worlds interpretation of quantum mechanics discussed briefly in Chapter 16, wave functions do not collapse but split into many branches. Some people argue that any observation of the universe is collapsing its wave function and is putting not only the observer but also the entire universe into one among many possible states. This is an extreme form of the many-worlds interpretation of quantum mechanics and is, not surprisingly, controversial.

What I now want to suggest, and already anticipated when emphasising the uniformity of the particle distributions in central configurations, is that all three of the effects for which inflation currently provides more or less persuasive explanations could be explained in a quite different way: purely classically through the royal zero that governs the universe at the big bang. I won't attempt to say anything about magnetic molecules, whose existence remains hypothetical.

Let's first consider the horizon problem. It exists as a perceived problem because the temperature of the universe is, to high accuracy, observed to be the same in all directions of the sky. In particular, if the standard FLRW models correctly describe the history of the universe, regions separated by only twice the moon's diameter on the sky could never have been in causal contact and equilibrated to a uniform temperature. Since the discovery of thermodynamics, uniformity of temperature has universally been attributed to causal equilibration; the observations therefore created a great puzzle. It disappears if, as a consequence of the law that governs it, the universe is necessarily born in a state that everywhere is close to thermal equilibrium and then very soon achieves that state by purely local interactions. Newtonian total explosions hint in that direction, as does the BKL conjecture, albeit with greater doubt because the corresponding theory is less developed. When discussing *N*-body total explosions in Chapter 16 I emphasised that the initial motions along the maximal eigendirection away from the very uniform shape at the central configuration will, with high probability, have a Maxwellian distribution of kinetic energies. Together with the further changes in the motions due to the effect of the other eigendirections, this will start to change the initial uniform state. Both the motions and the spatial distribution of the particles will look like they represent an equilibrium state. If the *N*-body behaviour also occurs in general relativity, full equilibration, with respect not only to kinetic energies but also to positions, will happen in exactly the same way everywhere in space. It's a direct consequence of the apparent conspiracy that a total explosion enforces. It has long been agreed that cosmological theory should predict that our universe satisfies the Copernican principle—when observed on large enough scales, it should look the same everywhere. The solutions with finite rather than zero size at the Janus point already led to such a prediction. The royal zero makes it much stronger.

Next, what about the flatness problem? If one takes the relational point of view seriously and argues the universe should be governed by a law of the greatest possible predictive strength subject to the sole condition that it lead to steady ongoing creation of structure, then in the *N*-body model the energy and angular momentum must both be zero. From the dynamical point of view flatness in general relativity corresponds to Newtonian solutions for which the energy is zero. In fact, before the theory of inflation came into existence, cosmologists were prepared to believe that some as yet undiscovered principle could dictate that only the special solutions corresponding to flat space should be realised. The ideas about the law of the universe developed in this book support such a proposal. However, a resolution of the flatness problem along these lines requires the universe to have infinite spatial extent and presents difficulties for the idea that it can be understood as a fully comprehensible holistic unity. An alternative, which I will discuss in Chapter 20, does allow the universe to increase in size without limit and become effectively flat.

The last thing to consider is whether there might be an alternative to the inflationary explanation of the Harrison-Zel'dovich spectrum through quantum fluctuations of an inflaton. As sketched earlier, this consists of two main processes—the emergence from a quantum soup of a classical spacetime, followed by exponential expansion accompanied with quantum effects which imprint macroscopic fluctuations on an originally smooth spacetime. Now, babies are born with an umbilical cord that is immediately cut, leaving a navel; it is not the case that at some time after birth navels appear on smooth tummies. Is it possible that, at its birth, the universe already had fluctuations of the kind that inflation is believed to have created after the big bang?

This, of course, is the possibility at which I have already hinted; in particular, it is implicit in the earlier putative resolution of the horizon problem. In fact, a possibility at least partially along these lines was considered in 2002 by Stefan Hollands and Robert Wald, whose paper with Joshua Schiffrin about a dart-throwing 'blindfolded creator' suggested the title of Chapter 14. Two distances play a controlling role in inflationary theory: the diameter of a causal patch and the wavelengths associated with possible quantum states. The key effect of inflation is that it greatly increases the quantum wavelengths with the consequence that at the end of inflation they extend well beyond the causal patch. If the effectively classical universe is born with such quantum fluctuations already in existence, inflation will be redundant. Hollands and Wald make only tentative suggestions how such a state could arise and admit that, from the quantum point of view, their proposal does not look as natural as inflation.

My alternative proposal is based on the appearance of the fluctuations in the particle distributions in Fig. 25, their similarity (at least to the eye) to the glassy distribution on the right of Fig. 24, and the assumption that the BKL conjecture is correct and that it has the consequence that the fluctuations in the cosmic fluid have a similar form and in fact a nearly scale-invariant Harrison-Zel'dovich power spectrum just this side of the big bang. In the absence of definitive conclusions in the BKL framework (the relevant conclusions that do exist are summarised in the notes to Chapter 17), one possibility is to see whether, for a very large number of particles (with more or less equal masses), the distribution of the particles in a central configuration with near-minimal complexity does not merely look like a Harrison-Zel'dovich distribution but actually is one. This would be intriguing and a definite hint that the cosmological fluctuations have a purely classical origin within general relativity and are nothing to do with quantum mechanics. Flavio Mercati and I did explore this possibility and discuss it with specialists, but with inconclusive results. One of the problems is that representation of the fluctuations of a continuous cosmic fluid by point particles can easily be misleading, especially as the Harrison-Zel'dovich distribution has a very precise nature. However, the hints from *N*-body theory do suggest an interesting line of research in general relativity. In particular, if it should turn out that the kinds of fluctuations needed to explain the large-scale structure of the universe are present in the vicinity of the big-bang spacelike hypersurface in classical quiescent solutions of general relativity, it would be a fundamental challenge to the widely held belief that quantum effects play a fundamental role in the origin of the large-scale structure in the universe.

Less dramatic and perhaps more plausible is that recognition of what the conditions at the big bang are like when represented in terms of the shape degrees of freedom (and the violent behaviour of the scale is ignored as gauge) might lead to a form of quantum gravity unlike anything currently expected, such that when a classical universe does emerge from a primordial quantum state it is already smooth and not chaotic. If so, the current difficulties with the initial conditions needed for inflation to get going might be alleviated. I think the main argument for this suggestion is that the curve representing the shape-space evolution in Fig. 23 becomes smoother, and not more complicated, at the big bang. Contrary to all the accounts one reads about conditions at the big bang, there might be a sense in which, as I suggested in Chapter 17, they are benign.

Let me end this part of the discussion with my justification for presenting the extensive material of this and the two previous chapters. I hope readers will have found the description of central configurations and the remarkable total collisions (with their alternative interpretation as total-explosion big bangs) of interest in their own right. However, the real reason they appear in the book is twofold: first, as an example of the form an actual past hypothesis (the extraordinarily special initial state of the universe postulated by Feynman, Penrose, and many others) might take together with the evidence that it could be a direct consequence of the law of the universe and not an 'add-on'; and, second, as an alternative to inflation, which is a pillar of modern cosmological theory and as yet has no serious rival—an unhealthy situation in any science, for theories must always be challenged.

In this connection I should like to point out that if the alternative to inflation suggested in this chapter on the basis of the material in Chapters 16 and 17 is on the right track, it looks as if it will be largely (indeed, perhaps entirely) free of the ad hoc elements in the theory of inflation. Despite its great success, it is often criticised for being a 'vanilla' theory. As I explained, there is nothing in the framework of the theory that determines the form of the inflaton; its properties are chosen to reproduce the observations. If it should be true that the Harrison-Zel'dovich fluctuations can be explained along the tentative lines I have suggested, it looks to me as if the theory will be to a very large degree unambiguous in the predictions it makes. We already see an indication of that in the particle model. Although the mass ratios cannot be fixed, the motions along the maximal eigendirection are completely fixed, together with the immediately following influence of the other eigendirections. This opens up the possibility that the initial shape is characterised by exactly scale-invariant Harrison-Zel'dovich fluctuations with unit slope of the power spectrum, which is then slightly reduced to the observed value by the first two kinds of motion described earlier. That would be a real advance.

I WILL END this chapter with further comment about laws of nature as contrasted with a law of the universe, specifically a quantum one. Maxwell, followed by Einstein, argued eloquently that they should be expressed through local differential laws. In following this principle they succeeded in eliminating the action-at-a-distance problem that had bothered Newton himself. In accordance with their field equations, causal effects cannot propagate faster than the speed of light. Since information can have causal effects, the same restriction applies to it. This is called Einstein causality and is very important in all modern attempts to understand the foundations of quantum mechanics. It would be a sensation if Einstein causality were to be overthrown, and I don't think it will be.

However, the elimination of action at a distance has had far-reaching consequences. They are best illustrated through their effect on the fundamental elements of dynamics: in particle dynamics the positions of the particles, their momenta, and the rates of change of the momenta (accelerations); in field theory the field values at different points of space, the rates of change of the field values (their momenta), and the rates of change of the momenta, which are determined by whatever forces are acting. Maxwell and Einstein eliminated accelerations generated instantaneously at a distance across empty space, but with a side effect that is striking but too little noted. Newtonian momenta, along with positions, can be freely specified at any initial time; this is not so in Maxwell and Einstein's theories or in modern gauge theories. Although individual momenta are not determined instantaneously by action at a distance, the complete set of momenta at any time are not freely specifiable but subject to constraints. The simplest example is the best-matching constraint of vanishing angular momentum in the particle model (Chapter 8). The constraints are *laws of the instant*; time plays no role in them. Four of Einstein's field equations, in many ways the most fundamental, are of this kind. For cognoscenti, they are elliptical, as opposed to the hyperbolic ones that govern the propagation of causal effects. They express interconnections of a universe that is holistic.

The material presented in Chapters 16 and 17 suggests there is still more to be unearthed. The laws of the instant constrain the momenta of subsystems of the universe, but if there is a true big-bang Janus point of zero size, there are much deeper interconnections in play. They constrain the initial data of the universe more tightly than any carpenter's vise. At the total explosion the momenta are completely controlled by the maximal eigendirection in a far more restrictive way than the vanishing of the angular momentum, the simplest nontrivial example of a gauge constraint. As regards the particle positions (or field values in BKL), their restrictions are unprecedented twice over: it is their first appearance in dynamics as theory, and it is also their debut in history—at the very birth of the universe. They promise cosmology's sine qua non—initial fluctuations that must be there and from which all the structures in the great wide universe can grow. If that were not enough, these binding laws are balanced by a freedom of an entirely unexpected kind. At the very point at which it appears the universe is in a straitjacket, the creation measure makes its decisive appearance, sets the cosmos on its way, and ordains how much creation there will be.

Is there a truly fundamental way to understand how that happens? I have said little about quantum gravity except that, contrary to expectations, it might take its simplest form at the Big Bang. Do shape-dynamic ideas give further hints? One thing is sure: the concept of time will be key. From the very creation of quantum mechanics, time has been the great enigma. It has a status quite different from dynamical quantities like position or momentum. They are 'quantised' and become operators with values determined only probabilistically and impossible to measure simultaneously. Time stays aloof, as absolute as Newton said (at least in the formalism, physical clocks are used to measure it in reality). One of the greatest problems in direct attempts to quantise gravity is that time disappears. The reason, reflected in the absence of d*t* in the Machian action in Fig. 7 (page 123), is that time does not exist independently of change but is abstracted from it.

We see that in Fig. 8 (page 127). There is no preexisting notion of duration, but successive configurations of the universe are 'stacked' in such a way that they do appear to evolve in a Newtonian absolute time. This is the emergent time of the universe. Kepler pairs also tell an emergent time as they become isolated at a sufficient distance from the Janus point and 'march in step' better and better with each other and the universe's time. If sufficient particles are present, another quantity advances quite closely in step with them: the complexity. The large fluctuations present in a three-body universe are smoothed in a 1000-body universe (Fig. 12, page 144); with enough particles present, the complexity will always increase. Might *it* be time? Fig. 12 can be interpreted in two ways: under evolution, either complexity tracks the emergent Newtonian time (taken to be fundamental) or it's the other way around and complexity literally is time. Suppose it is.

If so, here is a tentative sketch of quantum gravity. As suggested in Chapter 7, let shapes of the universe be instants of time. Let the lowest-complexity shape Alpha, α, be the origin of time, $t_\alpha = 0$. Then the time t of any other shape s is $t_s = C_s - C_\alpha$ and ranges from 0 to $+\infty$. All shapes with the same t_s are simultaneous. The shape potential being the negative of the complexity, time so defined measures the work done by the universe to reach the shape with time t_s. Quantum gravity will have a complex wave function Ψ defined on shape space. If shape space has n dimensions, the transverse subspace T_s of all shapes at the complexity time t_s has $n - 1$ dimensions. The probability of a shape s in the infinitesimal neutral-measure region ds around s will be proportional to $\Psi^*\Psi$. Shapes with a given complexity can have very different structures. Quantum shape dynamics will predict the probabilities for different structures with the same complexity—that is, the same amount of work done.

Schrödinger's wave equation for the complex wave function ψ has the form $i\hbar(\partial\psi/\partial t) = (-\hbar^2\Delta + \Phi)\psi$, with Δ the Laplacian defined on configuration space, Φ is the potential, and $\hbar$ has the dimensions of action. The quantum wave equation on shape space could have the similar but dimensionless form $i(\partial\Psi/\partial t_s) = (-\Delta_s + \Phi_s)\Psi$, where Δ_s and Φ_s are the Laplacian and shape potential on T_s. If this equation enforces unique WKB behaviour at Alpha (page 242) as the initial condition, the quantum state of the universe at all times will be determined. This proposal respects the spirit of shape dynamics. It's a tentative proposal, so I won't invoke it in the next chapter, but allow me to entertain conjecture of a time in shape space where it becomes manifest that such an equation explains why a certain shape is experienced and why Shakespeare wrote the things he did.

CHAPTER 19

THE CREATION OF STRUCTURE IN THE UNIVERSE

The Standard Account

It's time to bring all the conceptual and technical threads together and attempt to do justice to the *well-ordered cosmos*, a concept that we owe to Pythagoras. Entranced by the beauty of the world, he called it *cosmos*, the Greek word meaning primarily 'order' but also 'decoration, embellishment, or dress' (*cosmetic* has the same origin); in Latin we have *mundus muliebris*, the ornament of women. Thus, *cosmos* and *mundus* were conscious coinings to express the perfect order and arrangement of the world, perceived as an indissoluble unity. In the Renaissance this vision was a strong stimulus to the study of the celestial motions and the reawakening of the Platonic ideal of geometrical exactitude. Galileo's dissection of free fall to reveal its core in constancy of the acceleration—a new concept—is a gem.

It will be hard to match. In what I will call the standard account, I will present the best arguments I have found (with sources and some amplification in the notes) that take the second law to be inviolable and attempt to reconcile the strict entropy increase of the universe that it implies with the manifest structure it has created ever since the big bang. It certainly does not look like an increase of disorder; the mismatch calls for reconciliation. I will then contrast the standard account, which I find wanting in some respects, with the alternative Janus-point picture. Both rely on expansion of the universe, but I think the alternative takes it into account more directly, indeed almost visibly. I also think the concept of entropy in the standard account is problematic on two counts—it fails when the universe becomes inhomogeneous, and it accords tiny effects, which do increase entropy as it is often defined, unwarranted weight. I should say that the account in this chapter is, I believe, correct in essence up to our present epoch and, indeed, far into the future; the very far future is another matter and will be discussed in the final chapter. For alternative, more standard accounts of the whole of cosmic history, I recommend Sean Carroll's *From Eternity to Here* and Brian Greene's *Until the End of Time*. They take into account the widely assumed role of inflation, what is called the de Sitter horizon, and quantum fluctuations and tunneling. These may well be significant but are questioned by, among others, Roger Penrose in his *Fashion, Faith, and Fantasy*. In any case, I believe expansion of the universe, as long as it continues and leaves the ratios of relevant physical quantities finite, is the decisive consideration.

Meanwhile, I invite the reader to consider critically the extent to which these two chapters—and indeed the book taken as a whole—provide not only a good description of what we experience but above all a good explanation for *why* we experience it. In *The Beginning of Infinity: Explanations That Transform the World*, David Deutsch begins by setting out the criteria which genuine explanations, which need to work on many levels, must meet. An obvious one, necessary but not sufficient, is empirical adequacy: no predictions that a purported explanation entails must be contradicted by experiment. A far harder test, and the reason the history of science has seen so few transformative explanations, is that any proposed explanation of phenomena must simultaneously reproduce the established successes of earlier theories while making new predictions that are then confirmed by observations. It is very difficult to do both at once. This is why Einstein's general theory of relativity represents such a resounding success: it reproduced all the extraordinary achievements of Newton's theory, correctly predicted new effects, completely changed our view of the world, and, for those able to grasp the essence of its mathematics, did genuinely explain things.

The etymology of 'explain', from the Latin meaning 'spread out', makes it clear why good explanations give such satisfaction—we can literally see how and why things happen the way they do. Readers with insufficient mathematical knowledge and intuition to appreciate that in the case of Einstein's theory can see the effect much more readily in the Copernican revolution. We can all imagine ourselves looking down on a spread-out solar system: the planets, all more or less in one plane, circle the sun, with those nearer to it moving faster than those further away. From that vantage point, we can then easily understand why a terrestrial observer will see the planets making the strange motions that they do against the background of the sky. With his geometrical intuition, Galileo was an immediate convert. I hope the representation of big-bang total explosions through the topography of shape space as expressed by the eigendirections of the shape potential at a central configuration has a similar illuminating effect. Besides that, you must judge the extent to which this book's explanation of time's arrows, if it is correct, succeeds in the sense of laying bare the inner workings of the world.

As regards the prediction of new effects, the only ones I can see at the moment relate to the tentative proposal for the nature of the big bang. If the explanation I present here is correct, confirmation should eventually show up in the sky—to be precise, in the microwave background. For example, can the theory make an unambiguous prediction for the observed slight departure from unity of the fluctuation power spectrum and the point at which it turns over? Another thing that gives me hope that my collaborators and I might be on to something is that good explanations, as David characterises them, often come from a first simple idea that goes on to spawn many more—there is organic growth from an initial seed. That certainly happened with Copernicus's proposal: Kepler took it a great deal further with his discovery of the laws of planetary motion, and then Newton built a great edifice on them. The Janus-point idea has not yet yielded any new predictions, but it certainly grew from the simplest of seeds: Lagrange's 1772 discovery represented by the U-shaped curve on the left of Fig. 2 (page 96). All the ideas presented in this book, right through to the proposal for the nature of the big bang—what it looks like in shape space and its implications—came, one by one, from the initial realisation of the possible significance of that curve.

That's more than enough self-promotion. It's time to get on with what I have called the standard account. There are two main elements in the standard account: the notion of *free energy* and the definition of entropy in quantum mechanics. As mentioned briefly at the end of Chapter 1, free energy is a measure of the difference between the entropy a medium actually has at a given time and the maximum entropy that it could have. The difference means the medium can, before its entropy is maximised, do useful work and create structures. For an example, readers of an influential 2002 paper devoted to issues in cosmology are invited to consider a sealed box full of gas molecules in a low-entropy initial condition with all the molecules in a small volume in one corner of the box and so dense they form a fluid. Release of the molecules from the corner allows them to flow throughout the box, the process being described as follows: "For some time the gas is far from equilibrium. During this time, the second law insures that the entropy is increasing and interesting things can happen. For example, complex 'dissipative structures' such as eddy flows, vortices, or even life can form. Eventually the system reaches equilibrium, and all structures disappear. The system dies an entropy death".

Though the formation of life stretches credulity, the virtue of this simple example, which uses free energy implicitly, is the manifest increase of entropy directly coupled with the occurrence of interesting things. However, there are shortcomings. First, the initial condition is artificial and is what allows the molecules to expand freely in the early stages. Next, the initial free expansion will, by the role-reversal effect, give rise to attractor Hubble expansion, which is not particularly interesting; the situation changes only when the molecules encounter the second artificial feature, the walls of the box, without which eddy flows could not be set up. Finally, it is the box alone which allows the molecules to equilibrate and die an entropy death. The scenario bears little resemblance to what seems to be happening in the universe.

More persuasive are accounts that take into account known facts of cosmology and consider them explicitly in terms of free energy. They rely heavily on the entropic properties of black holes and estimates of the entropy in causally connected regions (defined as in Chapter 18) and comoving volumes, which are regions of space that expand at the same rate as the scale factor of the universe and can be defined in FLRW solutions due to the simplicity of their geometrical structure. Normal statistical-mechanical arguments can be applied in any given comoving volume because at any instant as many particles of each species leave as enter it. The conceptual box that played such a critical role in the discovery and development of thermodynamics reappears in cosmology—as the boundaries of the regions just mentioned and, in semipermeable form, as black-hole event horizons. A comoving 'box' certainly continues to play a useful role as long as the universe, with all the matter it contains, remains homogeneous and isotropic; the difficulty comes when that ceases to be the case. Moreover, while it might be argued that creation of structure is not incompatible with growth of an appropriately defined entropy, that is not quite the same as explaining how and why structures actually form, how they develop, and what they look like. This is what I will attempt later in this chapter.

The standard free-energy account accepts that just after the big bang the matter in the universe was in thermal equilibrium; therefore, entropy (in both causally connected regions and comoving volumes) was equal to the maximum possible. However, the expansion of the universe, while allowing entropy (measured, as we will see, by a count of the number of particles in existence) to increase, had the consequence that the actual entropy S_{actual} steadily fell behind the maximum S_{max} it could have. As a result, the ratio

$S_{\text{actual}}/S_{\text{max}}$ in both causally connected and comoving regions gets smaller and smaller. This means that the scope for free energy to do work can continue to increase for a very long time, and perhaps even forever. The underlying argument is that the creation of work and structure always requires a net increase of entropy, and in an expanding universe various processes allow that to happen. On the largest cosmic scale, the formation of black holes and their subsequent entropy-increasing evaporation are identified as the most important such processes. Structure formation can continue as long as black holes form. If, as seems likely, that cannot continue forever, an effective end state will be reached; it won't be one of literal heat death in thermal equilibrium but rather one in which structure ceases to increase and intelligent life cannot exist. However, various authors have considered ways by means of which life could be continued forever. These include engineering on a cosmic scale to bring about further entropy increase by changing the motion of black holes, thereby causing them to merge, with a consequent increase of the total entropy.

A first issue in the standard approach is its adoption (acknowledged to be without any first-principles derivation) of Penrose's conjecture that at the big bang the gravitational entropy of the universe was either very low or exactly zero. In particular, there should be no significant number of large black holes. That is at least clear in its significance and has the positive feature that, thanks to the work of Bekenstein and Hawking, there is a well-defined definition of black-hole entropy. However, Penrose's concept of entropy, beautifully illustrated by diagrams in all four of his books for the general public, is essentially classical and derives from Boltzmann. In contrast, the standard account is based very largely on the fact, which we will come to shortly, that in quantum theory the number of particles can change and that this change is the primary cause of entropy increase. It may also be wondered, given the unique position of the big bang in the history of the universe, whether *any* conventional entropy concepts can be expected to be valid there. In particular, the picture of a total-explosion big bang developed in Chapters 17 and 18 seems to bring with it a quite new form of entropy associated with the value of the emergent creation measure.

Another problem in the standard account is uncertainty in the value of the key quantity S_{max}; different ways of calculating the maximum possible entropy at any given epoch have led different authors to different values. It is also strange that although black holes, on account of their colossal entropy, play a crucial role in the argument that S_{max} is much greater than S_{actual}, black holes themselves do not seem to have had any role in the emergence of the quotidian arrows of time we find all around us. Indeed, profound unidirectionality of all processes throughout the universe is the defining characteristic of the early universe, when it seems no black holes existed; as we will see, that unidirectionality is entirely explicable as due to disequilibration of matter resulting from the Hubble expansion.

The final problematic aspect that I see in standard accounts relates not directly to the creation of structure but rather to the way entropy increase is treated quantum mechanically in cosmology. The basic principle is readily explained. You may have seen images of the explosive events that result from head-on collisions of particles accelerated to very high energies in the Large Hadron Collider in Geneva. The multiple particles created in such explosive events fly off in all directions, and the tracks they make provide clear examples of Thomson-type dissipation—the total energy is unchanged but what was previously concentrated in the two colliding particles is now distributed among many particles. Previously concentrated energy is spread out. Such processes can also be given a rough-and-ready entropic interpretation. In it, one simply counts the number of particles that exist before and after the collision and notes that the number, taken as a measure of entropy, has increased. It is a rather crude measure because it takes no account of the nature, energy, and distribution of the collision products. Nevertheless, cosmologists do often use a notion of entropy based simply on a count of the numbers of particles of different species.

There is a good justification for this. During the first few minutes after the big bang, huge numbers of transformations of particles into each other and into photons and neutrinos took place. The main issue of interest to cosmologists at the time of primordial nucleosynthesis, when the first nuclei were formed, is the number of photons per nucleon which existed at that epoch.* The number is large, 10^9, and is one of the most important in cosmology; it gives extremely valuable information about the early state of the universe. For this reason, entropy discussed in such terms plays a significant role in the theory of primordial nucleosynthesis. The simple count in terms of the number of photons per nucleon is, in fact, more precise than it seems, since virtually all the photons of interest at the time of nucleosynthesis were in thermodynamic equilibrium and thus had a very well-defined energy distribution, featured a uniform density distribution, and moved with equal probability in all directions.

However, this very uniformity in the epoch in which FLRW cosmological solutions are good approximations to the actual state of the universe highlights the shortcoming of the rough-and-ready entropic interpretation of particle collision and creation events. The mere count of the number of particles produced takes no account of the energies and species of those particles. It also does not in any way reflect the change in the nature of the particles' motions. Before the collision there are simply two particles approaching their common centre of mass along the line joining them; after the collision the produced particles fly off radially and more or less symmetrically in all directions. The change is 'invisible' in the FLRW solutions but it is steadily altering the state of the universe from near-perfect uniformity. It is sowing the seeds of the great structures we observe today.

Another difficulty with the count-of-particles definition of entropy has already been noted: it takes no account of the energies of the produced particles. Photons, the particles of light, had high

---

*Nucleons are the particles that form the nuclei of atoms; the two most important are protons and neutrons. The latter are stable when bound in a nucleus but when free decay in about ten minutes. This fact played a critical role in primordial nucleosynthesis. It is believed that protons too will decay given enough time, but very precise experiments have shown this must be immensely long.

energy in the early universe and existed in profusion; their number has barely changed since the CMB epoch but, due to the expansion of the universe, their energy has been greatly reduced. In fact, one of the most important results in cosmology—it was mentioned in Chapter 18—is that the energy density of the photons is inversely proportional to the fourth power of the time, whereas there is decrease only as the third power for matter particles such as protons and electrons. Because of this the energy density of radiation controlled the early expansion rate of the universe, but around the time at which the CMB came into existence that role passed to matter; there was a transition from expansion dominated by radiation to expansion dominated by matter. If entropy growth is such a defining characteristic of the universe's history, it seems odd that the energies of the counted particles should not be taken into account, like the root-mean-square length ℓ_{rms} and the complexity (as defined in Chapter 9) is mass-weighted to take into account the masses. The entropy-like quantity introduced in Chapter 15 is also mass-weighted because the state function which defines it is the mass-weighted complexity. If one does not take into account the energies of the particles counted, the calculated entropy of the universe can appear to depend to a bizarre amount on the slow cooling of planets and even rocks, which emit great numbers of very low-energy photons. I suspect we need as a state function the generalisation of the Yamabe invariant whose existence I conjectured and called Yamabe+Matter in Chapter 17. If there exists an invariant of what I called the universal group, it will be the perfect quantity to characterise the shape of the universe at any instant and serve as a state function to define an entropy-like quantity for the universe that, just as with the one defined for the particle model in Chapter 15 and illustrated in Fig. 16, will decrease with time, not increase.

Another problem related to attempts to define an entropy of the universe, as opposed to subsystems formed within it, is the great difference between the microhistories of confined systems and the unconfined universe. Except for fluctuations that are insignificant unless one waits for immense lengths of time and even then are unlikely to generate interesting structure, nothing remotely resembling a coherent history unfolds in confined microhistories. As we have seen, unconfined systems are quite different. They can create histories of great interest. We are denizens of one such system, the universe, and part of a monumental story that is still ongoing. Is entropy a helpful way to try to describe it? If, of necessity, we exclude from consideration dark matter and dark energy (whose true nature is still quite unknown, as we have seen), the great bulk of the mass/energy of the universe currently resides in stars and black holes. Both are in quasi equilibrium; in the case of stars we know pretty well what that state is like. The stars and the radiation they emit during the bulk of their lifetime have a basically spherically symmetric, onion-like structure in which the temperature and chemical composition vary smoothly with the distance from the centre to the surface in the case of the stars and into distant reaches of space in the case of the radiation they emit (the cooler radiation that the star emits earlier in its life is furthest from the star, just as it is for the radiation emitted by black holes). At any point within each star there is a state very close to thermal equilibrium, and the radiation emitted by any star at a given time can be well characterised by the temperature of the emitting surface.

But what are we to think about the distribution of all the radiation and matter—in stars, galaxies, and clusters of galaxies—in the universe at large? Does it have a well-defined entropy? Entropy is very often said to represent ignorance about the microstate of the considered system: a given macrostate has many microstates, but which one is realised at a given instant cannot be determined if one knows only the macrostate. Now, of course we have very little information about the state of individual particles in stars, but as seen in the summary just given, we do nevertheless know a remarkable amount about what is going on. In Chapter 1 I quoted Lamarck's comment that "the surface of the earth is its own historian"; exactly the same thing can be said about the current epoch of the universe. Its state proclaims its history and, as I said in Chapter 1, is an immensely rich and vast time capsule containing a multitude of mutually consistent records of the whole history of the universe more or less since the time of the big bang.

In 1965, Edwin Jaynes, who made many important contributions to probability theory, published a paper in which he drew attention to what he called the 'anthropomorphic' nature of entropy. He said that it might come as a shock to his readers that "thermodynamics knows of no such notion as 'the entropy of a physical system'". He pointed out that a thermodynamic system does have the concept of entropy but that the same physical system is associated with many different thermodynamic systems and corresponding entropies. In the terms that I used earlier, entropy is defined by a macrostate, and that can be defined with greater or lesser precision depending on the state variables used to characterise it. These characterisations, in turn, depend on the capabilities and choice of an experimentalist. Jaynes gives the example of a crystal of Rochelle salt, whose state at a given temperature can be defined in different ways that give different counts of microstates and hence different entropies.

Countering the suggestion that the problem arises because in any such choice not all the relevant degrees of freedom have been taken into account and that if they were the 'true' entropy would be determined, Jaynes points out that there is no end to the search for an ultimate 'true' entropy since it can end only when "we control the location of each atom independently. But just at that point the notion of entropy collapses, and we are no longer talking thermodynamics!" In fact, in cosmology the kind of problem that Jaynes points out is even greater because the very definition of mechanical systems becomes problematic. Where is the boundary of the sun? At its photosphere? That leaves out the chromosphere. Do the particles that truly escape from it into interstellar space still belong to the sun? And what about the radiation it has been sending into space for over 4 billion years? The fact is that the only possible candidate we have for a closed mechanical system is the universe itself. Of course, present cosmological knowledge leaves us very far short of Jaynes's location of each atom within it separately. There is still an important role for entropy in many places in the universe. But has the impressive growth of both theoretical and observational knowledge brought us to the point at which the search for an entropy of the universe ceases to be a helpful undertaking? The things we want to know are: Why does the universe seem to have begun in a big bang the way it did? What explains the unidirectionality of all macroscopic processes everywhere in the universe? Will it all come to an end, and if so, how? These questions are addressed in the remainder of this chapter.

The Emergence of Thermodynamic Behaviour

By emergence of thermodynamic behaviour I mean the coming into existence of subsystems of the universe whose instantaneous states and behaviour, including entropy increase, suggested the idealised foundations of phenomenological thermodynamics and statistical mechanics. The distinction between the universe and its subsystems is critical; so far as I know, it has not hitherto been clearly made in the discussion of time's arrows.

I begin by expanding on the comments made at the end of Chapter 15—namely, how the decrease of the entaxy of the *N*-body model with increasing distance from the Janus point reflects formation of self-confined subsystems of the universe. I then discuss the manner in which the subsystems, as long as they exist, behave like thermodynamic systems. The phrase 'manner in which' is important. As I have emphasised, virtually all discussions of entropy and thermodynamics presuppose, often unconsciously, systems that are subject to either perfect confinement or effective confinement (in the case of a system that can exchange particles with a thermal bath). But that is artificial. Such systems are found only in theoretical papers published in scientific journals. In the rest of the universe, confinement is always imperfect and transient (even if it does exist in effect for billions of years in the longer-lived stars). The formation of quasi-stable systems followed by their decay is the norm.

These are not only the systems with which practical applications of thermodynamics and statistical mechanics deal; they are also the systems whose behaviour defines for us a direction of time. If we can understand how they all arise and decay irreversibly in the same temporal direction, we will have cracked the problem of time's arrows. The all-universe perspective is essential and allows description in terms of both Thomson's energy dissipation and Clausius's entropy increase. However, the invariable association of entropy increase with increasing disorder, which is certainly true for confined systems within the universe, can be challenged in the context of the universe. The emergence of a rod, clock, and compass in the three-body problem and, on a grander scale, Newton's 'most beautiful system' provide ample grounds for doubt.

In formal mathematical terms the central contention of this book is that the mystery of time's arrows has, up to now, remained unresolved because of the assumption, either explicit or implicit, that the phase space of the system under consideration has a bounded measure—the system is confined to a box. Zermelo's invocation of the recurrence theorem was fatal: Boltzmann was unable to advance a definitive explanation of the entropic arrow and was forced to seek refuge in rare fluctuations in a universe whose natural state is one of thermal equilibrium. The entropic arrow we experience could then be explained by the fact that we live in such a fluctuation as it emerges from an entropy dip. This idea was a lasting achievement—it marked the end of the Newtonian concept of absolute time that flows uniformly forwards quite independently of what is happening in the universe. Instead we recognise the direction of time in the nature of the change we observe around us. However, the fluctuation idea could not explain satisfactorily the evidence that we live in the huge evolving region that Boltzmann envisaged. The chances of a fluctuation creating such a region are negligible compared with the transient creation of a solitary brain deluded into thinking it lives in such a region. Eddington in 1931 and Bronstein and Landau in 1933 were the first to see the 'Boltzmann brain' flaw (Chapter 5).

Boltzmann, however, had noted almost in passing that the recurrence theorem would fail if the system considered is the universe and does not have a bounded phase space. Very understandably, given the complete lack of useful cosmological information, he made no attempt to explore possible implications. What I find remarkable, and highlighted in Chapter 5, is that the critical role of the recurrence theorem and the premise needed for its proof have hardly featured, so far as I have been able to find, in the discussions of time's arrow since expansion of the universe was discovered. But, as we have seen, the very first qualitative result in dynamics—Lagrange's in 1772—provides the simplest nontrivial example of a system, the three-body problem with non-negative energy, whose size increases without limit in both time directions. It does not have a bounded phase space. This is also typically true of systems with negative energy and more than two particles. They generally do not have a unique instant of minimum size, but they do typically break up and disperse on either side of a 'Janus region' of the kind I described in Chapter 6 when I introduced the notion of the Janus point.

If a maximal-entropy state like thermal equilibrium is not the natural state to be expected for the universe throughout its history, the previous chapters do suggest it is likely to have had a spatially uniform shape at the Janus point, especially if there the size of the universe was zero. Whatever may be the case, the role-reversal implication of Liouville's theorem for the shape degrees of freedom in an expanding universe shows that a pervasive arrow of time from higher to lower uniformity must be manifested throughout the universe on either side of the Janus point. The aim now is to spell out how the arrow is actually manifested. This requires inclusion of all forces of nature, general relativity, and quantum effects. However, the *N*-body model already illustrates very simply certain key effects without the complications they introduce, so it is still worth considering what further insights it gives. Moreover, it shares with the far more realistic modern cosmological models the two architectonic features that this book identifies as responsible for time's arrows: a dynamical law which guarantees fulfilment of Liouville's theorem and expansion of the universe. This suggests that, viewed in the appropriate context, there is nothing remarkable about the arrows. They are an inherent feature.

THUS, LET US explore a little further the implications of the *N*-body model, bearing it mind that its failure to satisfy the Copernican principle does not diminish its value if there is a large region around its distinguished centre of mass in which the conditions are sufficiently uniform. Such a region, the existence of which can be expected plausibly on the blindfolded-creator argument and with certainty in the total-explosion situation, will then model the conditions in a universe in which the Copernican principle is satisfied. An important feature of the *N*-body model, which it shares with any universe that has a Janus point, is that the evolution away from that point takes place in accordance with the same laws everywhere. Since the evolution in the region that models the universe begins with similar conditions everywhere, this has the potential to explain why throughout the universe we observe localised processes that all define the same direction of time. The argument is very simple: because all regions begin looking the same and are subject to the same laws of evolution, they must therefore continue to look the same. This can resolve, at a stroke, what has long been seen as one of the great problems with time's arrows: why do they all point in the same direction?

The puzzle exists as long as one supposes that the subsystems in which the arrows arise are independent of each other. Under this assumption, given the time-reversal symmetry of the dynamical laws that govern the subsystems, there does not appear to be any explanation for the remarkable alignment of all the arrows. However, in a properly enlarged perspective, the law of the universe necessarily creates, with high probability, a prior condition that enforces identical subsequent behaviour on all the subsystems. It is true that, once properly formed, they are dynamically independent in that they do not exert significant forces on each other, but they are not independent as regards the manner of their birth. In scientific studies, especially in biology, it is a golden rule to avoid the assumption that an observed correlation between two effects implies that one is the cause of the other—two correlated effects may have a common cause. This, I suggest, is the origin of the unidirectionality of time's arrows in what superficially appear to be independent systems.

The creation of effectively isolated subsystems has further important effects. As we have seen with Kepler pairs, the subsystems serve as rods and clocks. They can be used to determine dimensionful quantities. We can start to talk about temperatures and differences in temperature between subsystems. Moreover, the subsystems are, to a greater or lesser extent and for shorter or longer times, 'self-confined'. As long as that state persists, they 'make their own box' and have a bounded phase space. This means that the standard tools of statistical mechanics can be applied to them. In a beautiful account Brian Greene demonstrates this very effectively for stars in *Until the End of Time*. Once such subsystems are sufficiently isolated, entropy can, to a good approximation, be defined for them as a meaningful dimensional concept that exists alongside the universe's dimensionless and decreasing entaxy. As we will see, the overwhelming majority of the subsystems will have an entropy that increases in the same direction. This all comes about because attractors act on shape space.

In fact, they act not just once on the universe as a whole but also on the subsystems that form. Indeed, as they become ever more isolated in the expanding universe, they exist, when properly formed, in effectively empty space. Unlike a Janus-point universe with non-negative energy, they have negative energy; they could not be held together otherwise. However, sooner or later the great majority of them will break up and their parts will disperse into space. This will not be such a clean process as it was for the universe since, as we have seen, the subsystems will have a range of possible energies and angular momenta, in contrast to the vanishing values for the universe as whole. Moreover, they will have not a unique Janus point but a 'Janus region'. This does not affect the fundamental point: being unconfined systems with a size that in general can increase without limit, they will, just like the whole universe, be subject to attractors on their shape spaces if, as is typical, they do expand.

The big picture is this. The universe as whole has a uniquely defined Janus point and bidirectional arrows of time. The observable dynamics of the universe immediately begins to be time asymmetric in either direction from the Janus point (although initially only weakly). As regards emergent subsystems, their very existence is not always clear, and when they do exist, their boundaries in both time and space are much harder to pin down. However, what cannot be doubted is that subsystems, defined as a matter of pragmatic judgement, do come into existence. For who would say the sun and moon do not exist? Moreover, the mere fact that essentially all the subsystems of macroscopic size are born and then decay gives them an arrow of time that exists as long as they do. Unlike the universe with its bidirectional arrows, the subsystems have monodirectional arrows, all aligned with each other and with the master arrow dictated by the expansion of the universe on their side of the Janus point.

What is more, there are in general subsystems within subsystems, each and all with their own somewhat indefinite 'birth dates' and monodirectional arrows aligned with the master arrow. In fact, an extended hierarchy stretches from individual unstable elementary particles and radioactive nuclei at one end right up to clusters of galaxies at the other. To justify this statement, the next section will describe in more detail how the *N*-body picture just summarised matches the picture that cosmologists currently have of the successive formation of structured subsystems in the universe. It will have the advantage that it takes into account both quantum effects and Einstein's general theory of relativity. It will also describe the interesting structures and motions that simply cannot arise in permanently confined systems.

Birth and Death of Interesting Things

In this part of the chapter, I will allow myself some liberties. I think this may be permitted since the very earliest stages in the evolution of the universe are subject to considerable uncertainty. I am going to suggest that when it comes to understanding in detail the innumerable forms that time's arrows take in the great variety of systems that come into existence in the universe through the quantum mechanically governed action of the various forces of nature, the relevant effects can be characterised as *clumping* and *spangling*. By clumping I simply mean the formation of structures that come into existence and are held together as a collection of particles; their number can be anything from two or three up to the vast number in a galaxy. Clumping increases the complexity of the universe, doing so in a mass-weighted manner, as I emphasised in the first part of this chapter. By spangling I mean 'firework' effects like those created by the collisions of particles at the Large Hadron Collider.* Clumping will occur simultaneously if the collision results in not only a firework but also a bound system of particles (what I will call a bead). At much lower energies than in the first seconds after the big bang, the kinds of process I mean happen when a proton captures an electron in one of the higher energy levels of the atom that is thereby formed, after which one or more photons are emitted as the electron falls to its ground state. The resulting stable atom is the bead, the photons the firework. The bound systems formed in the early universe are almost all unstable and decay rapidly in decay products. The beads spangle.

---

*A spangle, or sequin, is "a small thin piece of glittering material, typically sewn as one of many on clothing for decoration"; as a transitive verb, 'to spangle' means to "cover with spangles or other small sparkling objects". I will also use 'spangle' as an intransitive verb; in the example I gave, the collision stimulates the colliding particles to spangle through the creation of the outgoing particles. Those of us lucky enough to have dark skies at night can always appreciate the star-spangled sky. It is not clear to me from the account in Wikipedia whether it was Mary Pickersgill, the creator of the original Star-Spangled Banner, or the commissioning commander at Baltimore, George Armistead, who conceived the design of what was to become the national flag of the United States; clearly stars were the inspiration. My poetic language may be raising the eyebrows of sober scientists; I use it to make more vivid the difference between what happens in confined thermodynamic systems, on the one hand, and the kinds of motions and structures that exist and can be observed in the universe, on the other.

I now argue that both clumping and spangling are processes that increase order, understood as the creation of interesting structures and motions, even though they simultaneously increase entropy as it is usually defined. If this seems oxymoronic, consider spangling. In the rough-and-ready sense, making no allowance for the energy and species of the produced particles, spangling does increase entropy, but does it increase disorder? The sparks that fly from a firework—say, a Roman candle—are anything but disordered; they would not give a fraction of the pleasure they do if they were. In contrast, peas ejected simultaneously in all directions from a small box within a large box would, despite initial regular motion, soon be bouncing all around it in a disorderly way. Just as Thomson said, dissipation of mechanical energy is universal, but he could have called it *spreading*; when that occurs in an unconfined situation the result is not disorder but order.

There are no eternally confined systems in the universe. Carnot brought them into physics with his idealised steam engine. Thermodynamics and statistical physics are mathematical constructs admirably chosen to give an approximate description of locally observed phenomena. But the box is a fiction; it does violence to the proper conception of nature and the universe. Copernicus said (page 172) of his predecessors that their description of the motions of each planet separately had made it impossible for them to deduce "the structure of the universe and the true symmetry of its parts". An artist taking from various places hands, feet, a head, and other bodily pieces would not create "a representation of a single person but rather a monster". For 'person' read 'the universe' and you have the problem with thermodynamics. Its plethora of experiments, all with conditions under close control in a box, blinds us to the truth. We see a monster: the second law. At the time of writing COVID-19 is on the rampage around the world. Eddington's warning of deepest humiliation for anyone whose theory questions the second law equates the law to a virus for which a vaccine will never be found; it will forever spread disorder near and far throughout the universe. Newton, with a boy's eye, knew that much lay undiscovered; he glimpsed it in the solar system. The well-ordered cosmos, not disorder, is the work of time. Clausius got it wrong; the complexity of the universe tends to a maximum.

I need an example to make the point. Almost the only interesting things that happen in thermodynamics as conventionally taught to students are phase transitions, the complete changes of state such as when water vapour in a container condenses and the water then freezes. In laboratories such processes are investigated under quantitative control by letting the studied system lose energy as heat flows out through the container walls into cold ambient air. Loss of energy allows the vapour to condense into water and the water to freeze. Ice can acquire fairy-tale structures. With their sixfold symmetry snowflakes are always a delight, but no two are ever exactly alike; it's the lack of perfect symmetry which makes them the beautiful things that they are. Have a look at some online.

Why is it that snowflakes can form? To feel the reason in the cold of a winter's night, put water in a bowl outside your door to let it freeze. Or stay indoors, read Shakespeare, and grasp the reason in your mind:

When icicles hang by the wall,
And Dick the shepherd blows his nail,
And Tom bears logs into the hall,
And milk comes frozen home in pail,
When blood is nipped, and ways be foul,
Then nightly sings the staring owl,
To-whoo;
To-whit, to-whoo, a merry note,
While greasy Joan doth keel the pot.

It's not the owl that causes the icicles to form. It's the expansion of the universe that, almost from the dawn of time, has been diluting and cooling the radiation that bathes the universe and, with it, the earth. The dilution is why the ratio of the temperature of interstellar space to the air by your house or apartment door is much less than 1 and why heat flows from the water and allows it to freeze, along with the icicles that hang by the wall. On less cold but

still clear nights water vapour condenses as dew droplets on grass.*
Without them we would not have in *Hamlet* the lines

> But look, the morn, in russet mantle clad,
> Walks o'er the dew of yon high eastward hill.

The dark sky at night was seen, in one way or another, as a puzzle by several people in the past, including Kepler; Edmund Halley, who predicted the return of the comet that bears his name; and, in the 1820s, the physician and amateur astronomer Heinrich Olbers, from whom Olbers's paradox gets its name. The paradox arises from the assumption that the universe has infinite size and is filled throughout with stars like the sun. Whatever the distance between them, the night sky should look as bright as the sun by day, but of course it doesn't. Hubble's discovery resolved the paradox. The main reason Boltzmann had no chance to explain entropy growth was that he had no inkling the universe could be expanding. Was he aware of Olbers's paradox? Thomson, by then ennobled as Lord Kelvin, was, and in 1901 he said its only possible resolution would be "by supposing the distance of the [conjected but invisible stellar background] so immense that no ray from it has yet been able to reach us at all".

Although he was quite wrong with that idea as well as on the age of the earth, Thomson was right when in 1852 he said there is universal dissipation of mechanical energy. Paul Davies recently called this "one of the gloomiest scientific predictions of all times" and said that one "need look no further than the Sun, slowly burning through its stock of nuclear fuel, radiating heat and light irreversibly into the cold depths of space, to see an infinitesimal contribution to the approaching heat death". When discussing the paper of

---

*In his *The Lightness of Being*, Frank Wilczek comments that atoms are generally found in their simplest form, their ground state, because they "are starved for energy", for which the ultimate reason is that "the Universe is big, cold, and expanding. Atoms can pass from one pattern to another by emitting light, and losing energy, or absorbing light, and gaining energy. If emission and absorption were balanced, many patterns would be in play. That's what would happen in a hot, closed system".

1852, I said a complete law of things should explain how the stores whose energies are universally dissipated arise in the first place. That question has already been addressed to some extent in the book and is a main topic in this chapter. Any such stores must each, almost by definition, exist within a confined region. This rules out the idea that the universe itself is the ultimate store. It is likened better to a warehouse within which clusters of matter, each a store, are distributed. We see the simplest—archetypical—example of their incipient formation in the way the singleton and the Kepler pair emerge from the Janus point of the minimal model. The law of the universe allows them to become separate: the singleton with purely positive kinetic energy, the pair with kinetic-potential deficit to balance it. Neither is yet an energy store; for that more particles are needed. The discussion in the second part of this chapter, as already earlier in the book, showed how particles can in arbitrary number be bound in cluster form but then escape into celestial space. Each escape is an event in which energy is dissipated, and the amount that remains in the cluster has been reduced.

NOW WE COME to the promised description of the birth and death of interesting things. The transformation of the universe from its initial state to what we observe now has already been covered in earlier chapters: the Janus-point conjecture in either the form initially discussed or in the radical total-explosion form of the three previous chapters with their tentative proposal for the shape of the big bang and the origin of the Harrison-Zel'dovich fluctuations. I will begin with the state that results, in the widely adopted inflationary account, at the end of 'reheating'. It is a soup of particles that, apart from the scale-invariant Harrison-Zel'dovich fluctuations, has an extremely uniform density and isotropic distribution of directions. The universe, with its dynamical evolution governed by radiation (much of it in the form of photons, the particles of light), is in near-perfect thermal equilibrium at an extremely high temperature and, in the standard account with scale, expanding very rapidly. This is what allows interesting things to happen.

As yet there are no beads in the soup, but then commence clumping and spangling, the two main processes which I argue are responsible for the continual change in the structure of the universe and essentially its whole story. The first significant thing is the formation of baryons, which are heavy particles like the proton and neutron; the electron, which is nearly 2000 times lighter than either, is a lepton. The creation of the baryons will typically be accompanied by the formation of other particles that fly off in 'mini-spangles'. Although any description of quantum events and states in classical terms is problematic, I think it is reasonable to liken baryons to the more or less stable clusters that form in the N-body model. The important thing about the creation of the first baryons and any accompanying spangles is that the universe begins to acquire structure. As I suggested (page 145) when discussing the notion of complexity, its use is not restricted to gravitationally induced clustering or clumping. In principle, some suitable definition of complexity—the putative Yamabe+Matter invariant of Chapter 17—can be used to characterise any non-uniform matter distribution. Accordingly, the creation of baryons will increase the complexity of the universe.

Their formation also represents what looks to me like the first clear example in actual cosmic history (as opposed to the illustrated N-body model) of the coming into existence of one of Thomson's stores of mechanical energy, for in accordance with Einstein's $E = mc^2$ each baryon contains a great amount of energy. Moreover, with the exception of protons, all baryons soon decay: neutrons within minutes, others much sooner. They are therefore excellent examples of the dissipation of mechanical energy. What we now have, which Thomson did not, is a potentially complete explanation of the unidirectionality of all processes (except the fully comprehensible violations of entropy increase at the microscopic level) in the universe. This includes not only the irreversible dissipation of energy stores but also their coming into existence, sometimes through the decay of previously existing stores. The explanation is potentially complete because the need for their previous existence does not involve an infinite regress—everything starts at the Janus point.

Moreover, the state there, although special, is not improbable, inexplicable, and in need of an extra condition to be imposed upon the law that governs the universe. Quite the opposite, that law, at least one of the kind assumed in this book, dictates the special state, which must be more uniform than all which follow it on either side of the Janus point. Contrary to what Feynman thought when he gave his 1964 lectures at Cornell University, it is not "necessary to add to the physical laws the hypothesis that in the past the universe was more ordered, in the technical sense, than it is today". For no hypothesis of uniformity needs to be added to the laws. The Janus point must exist, and the laws dictate uniformity at it either on account of the condition that must be satisfied if the universe has zero size or simply because a uniform state is more probable in line with Boltzmann's insight, transferred to shape space, that there are many more uniform states than non-uniform ones.

From the formation of baryons on, there follows the creation of the whole hierarchy of structures that I mentioned at the end of the previous section. It will help to take you through the outstanding stages of the creation process, not only for its own sake but also to characterise the universe in terms appropriate for an unconfined system. In it the motions and structures are quite different from those in a confined system. The difference is true both for the universe in its entirety and for subsystems within it. As noted earlier, expansion of the universe makes it possible for subsystems that, on each side of the Janus point, become effectively isolated, to behave like the whole universe and to exhibit attractor motions. The motions associated with spangling are precisely such and underline the dichotomy between microhistories of confined and unconfined systems, the former doomed forever to wander from one disordered state to another while the latter thread their way delicately along distinguished paths through phase space to the ultimate destinations to which they are guided by the universal attractors on shape space. Along the way to those destinations, whose nature we can at best guess, the microhistories exhibit wonderful effects.

Spangling, like Hubble expansion, is very special. We need the concepts of clumping and spangling, which are the very opposite of what happens as confined systems equilibrate, to properly characterise and appreciate the most important irreversible processes that take place in the universe and determine for us the direction of time.

After these preparatory comments, I can now briefly chronicle a selection of the whole gamut of subsystems that, in the course of the universe's history, have come into existence, evolved, and perhaps already decayed, though many, like the sun and earth, still exist. In the first three minutes after the big bang, nuclei of helium were created along with protons (the nuclei of hydrogen) and a few other of the lightest nuclei in the process of primordial nucleosynthesis I mentioned earlier. Steven Weinberg tells this part of the story superbly in his book *The First Three Minutes*. As yet, no electrons had been bound to the nuclei to form permanent atoms because the temperature was too high; for about 300 000 years collisions with the background photons almost instantly destroyed any that did form. Then, as the universe expanded, it 'stretched' the wavelength of the background radiation and steadily reduced its intensity to the point at which electrons could remain bound to the nuclei to form the first atoms (almost exclusively of hydrogen and helium), the universe became transparent, and the photons streamed freely through space; apart from a large increase in their wavelength, which grows as the universe expands, they are observed in the current epoch essentially unchanged as the CMB.

The creation of the first atoms was followed by the 'dark ages', the period believed to have lasted for several hundred million years in which the universe was completely dark until the first stars were born. By that time the appearance and structure of the universe were greatly changed from the first three minutes, when it was still to a very high degree isotropic and homogeneous. I said not only appearance but also structure because, according to current theory, it was during the dark ages that dark matter began to have a very significant effect on the evolution and structure of the universe. Through its self-gravitation the dark matter formed haloes within which ordinary matter (largely consisting of baryons) began to collect and form stars and galaxies. The two effects together therefore resulted in the formation of gravitational 'clumps' at two levels. First were the galaxy-hosting clumps of dark matter; of great spatial size, these clumps were separated from each other by increasing distances due to the expansion of the universe. Second were the stars within the galaxies. Before the onset of the large-scale clumping, the initial soup had merely become studded throughout by 'beads' in the form of nucleons, which by the time of the epoch when the CMB is observed had become hydrogen and helium atoms.

The first stars consisted almost entirely of hydrogen and helium, and they initially burned hydrogen to create more helium. This process, which Brian Greene describes so well, generated surplus energy, most of which left the stars in the form of the radiation which enables us to see them. Besides the clumping due to gravity that created stars (and still does), further clumping started to take place at the microscopic level through the creation in the stars of more massive nuclei from less massive ones. Meanwhile, the stars began to spangle as they radiated energy into space, more or less uniformly in all directions. The universe became a fireworks display.

Although it is undoubtedly right from the point of view of statistical mechanics, this prompts me to a certain disagreement with Greene's account of what happens in stars, which he likens to the action of steam engines and their need to 'shed entropy' by dumping heat since otherwise they would seize up. I see two differences. First, stars do not run through a potentially endless number of identical Carnot cycles; their state continuously changes as they create in their interior nuclei with increasingly rich structure (I will say more about that in a moment). Second, a steam engine dumps its waste entropy into what is an effectively confined and more or less equilibrated medium, which it simply heats. In contrast, the stars shed their heat by radiation into empty space—like fireworks, they spangle, which is why we, light-years away from them, see them sparkle in the sky. This is the difference between stars, which are open systems, and closed systems. The steam from a hot shower spreads throughout the bathroom and there is effective equilibration. To truly appreciate what is happening in the universe, we need to consider not only the details of what happens in an individual star but how the processes in all stars change the appearance of the universe that has brought them into existence. In fact, the action of steam engines should be interpreted similarly—by lifting coal from mines to the surface of the earth, the early exemplars, like stars, were doing their bit to change the shape of the universe. The true significance of the second law of thermodynamics needs to be interpreted in those terms.

To continue with the story of the universe: During the 13 billion years or so since the formation of the first stars, ever heavier and more complex nuclei, up to iron, have been 'cooked' in the stars by nuclear fusion. After iron, though, much more violent processes than the relatively mild fusion that takes place in stars are needed to continue the work of creation. It is here that the mighty supernova explosions take over and make the heavier elements. When one happens, a single supernova can briefly—for days or weeks—outshine the entire galaxy that hosts it. It also ejects vast numbers of atoms, both lighter and heavier than iron, into space.

One of the great insights we owe to the understanding of stellar evolution and the role of nuclear reactions in it is that we are literally made of stardust, mostly in the form of atoms up to and including iron. But we also decorate ourselves with the heavier elements, above all silver and gold created in supernova. They are called noble elements because, unlike iron, they do not rust and decay. But their nobility also reflects the grandeur of the whole process of the universe's history that brings them into existence. When Paul Davies called Thomson's prediction one of the gloomiest of all times in science, was he not adopting the specific anthropocentric view that goes back to Carnot's aim of optimising the use of coal for the immediate aims of humankind? From that parochial point of view the sun's radiation of the overwhelming bulk of its stock of nuclear energy into the cold depths of space does look like an absolute waste of mechanical energy. But considered from the point of view of what is truly important for humankind—and perhaps the universe—without it we would not be here at all, able, following Carnot's example, to optimise the use of whatever energy is available to us and to delight in a multitude of things. Without the profligacy of innumerable stars, all doing what our sun does, there would be no stardust. The labour of the stars is an integral contribution to a story of ever-increasing complexity and the creation of the most beautiful structures imaginable. It is this, not growth of disorder, that puts the direction into time—and us into the universe to witness its forwards march.

Time presents, indeed is, a story of great drama: supernova explosions of sufficiently heavy stars lead to the most dramatic events in the cosmos, the formation of black holes. This is clumping on the grandest and most concentrated scale; a colossal quantity of mass and energy collapses into a relatively very small region—a mile across for a star with the mass of the sun. By any measure this greatly increases the complexity (clustering of matter) of the universe. It also results in spangling because gravitational waves—ripples in space—spread out radially in all directions. Just as we see stars with our eyes and the Hubble Space Telescope sees galaxies with its mirror, gravitational wave detectors have, since the first triumph in 2015, been 'seeing' the merging and creation of black holes.

Like me, you may have watched the news conference that gave details of the detection of the waves generated by the merging of two black holes, which had masses about 36 and 29 times that of the sun. The simulation of the event as the two progenitor monsters spiralled ever closer to their destruction and the birth of an even bigger beast was spellbinding. So too was the subsequent depiction of the gravitational waves radiated into space. It was just like watching the effect of a stone thrown into a still pond. But the difference in energy is hard to grasp. The mass of the newly created black hole was not 36 + 29 = 65 solar masses but only 62. Accordingly, the gravitational waves carried away energy equivalent to the transformation of three suns into pure radiation in accordance with $E = mc^2$. Once again, that's a demonstration of the dissipation of mechanical energy, albeit on a scale Thomson could hardly have conceived. It's also a mighty example of spangling, though actually no more beautiful to watch than the waves the stone creates in the pond.

WAVES IN A pond are another of time's arrows that I mentioned at the start of Chapter 1. They are said to be retarded because they are observed after the effect—the stone thrown by a child into the pond—that causes them. They are manifested so clearly because the pond is flat and tranquil before the stone is thrown. They are regarded by some as a mystery because the laws of fluid dynamics are time-reversal symmetric and perfectly consistent with the existence of coherently synchronised excitation all around the pond's bank which initiates a circular advanced wave motion that converges on the centre of the pond and ejects the stone back into the hand of the child. Just at the start of the Second World War, John Wheeler and his recently recruited student Richard Feynman began to think about the analogous issue that exists in the transmission of radio waves. How does it come about that they always leave the transmitter and never converge on it? In accordance with Maxwell's theory of electromagnetic waves, nothing in nature should prevent that. Wheeler and Feynman developed an elegant theory in accordance with which any electric charge must, if it undergoes acceleration induced by any force, emit radiation that is precisely symmetric: half is emitted forwards in time and is hence retarded and half is emitted backwards in time and is said to be advanced. That only retarded radiation is observed in the world is attributed to the universe as a whole. The radiation emitted forwards is assumed to undergo complete absorption by the charges of the universe in the future; this causes these particles, collectively called the absorber, to emit radiation half forwards and half backwards in time, which travels back to the original emitter. The radiation that the charge emits backwards affects charges in the past, which also emit symmetrically. What they emit forwards cancels what the particles of the absorber in the future emit backwards, and the net effect is pure retarded emission by the considered charge.

Wheeler and Feynman point out a serious difficulty in their theory because it is "from the beginning symmetrical with respect to the interchange of past and future".* Why does an oscillating charge in an absorbing medium invariably lose energy? They point out that this happens not because there is complete absorption in the future but rather because they had built asymmetry into their theory through the improbability of the initial conditions. While no objection can be made to such a solution of the equations of motion, they had in effect proposed an initial condition like the one that can always be imagined in statistical mechanics by simply exactly reversing all particles at the end of an equilibration process. That will make them evolve to a highly improbable non-equilibrium state. Wheeler and Feynman's proposed law is, I suggest, an awkward mixture of local and global. The half-advanced–half-retarded emission hypothesis governs a local event supposed to take place in spacetime under all conditions quite independent of the state of the universe at large; it's a local law of nature. In contrast, the role of absorption brings in a Machian element—the universe at large influences what happens locally.

In the Janus-point scenario, and indeed in accordance with modern cosmology, the universe was in near-perfect thermodynamic equilibrium at least very soon after the big bang. There was only the slightest trace of time asymmetry in the phenomena that unfolded; the expansion-induced attractors had only just begun to exert an effect. In thermal equilibrium there is no distinguished arrow of time of any kind, neither entropic nor electromagnetic. Such arrows can appear only once matter has begun to form clumps and space opens up between them. Then particles can escape from the clumps into the surrounding space and Thomson-type dissipation takes place. As regards the emission of electromagnetic waves, the different behaviour of matter and radiation in the expanding universe is decisive. I mentioned that the energy density of radiation decreases inversely as the fourth power of the time, whereas the matter density decreases inversely only as the third power. That, as we have seen, is why water in a bowl by your back door freezes

*Despite intermittent interest in the absorber theory, I am not aware of any consistent form of it.

overnight and icicles hang by the wall in the morning. The space between the regions in which matter collects is effectively swept clean of radiation by the expansion of the universe. It's the counterpart of the tranquil surface of the pond. Only the most highly synchronised conditions in the thermal bath of the early universe could evolve into radiation converging on a particle that is about to be accelerated.

In summary, I think expansion of the universe explains all retarded waves observed in the universe, including those with the colossal energy generated by merging black holes, a child's stone thrown into a pond, and the waves of even lower energy broadcast to the world by radio and TV transmitters. Through the attractor mechanism it also creates localised subsystems of the universe governed (though only to increasing accuracy) by time-symmetric local laws that should be seen as emergent, not the fundamental building bricks of the universe.

IT'S WE HUMANS who see the stars and watch so many beautiful effects, delight in music and read poetry. That we can do so relies on my final example of clumping and spangling. Astronomers now know that the formation of planets around stars is ubiquitous. Very delicate observations have already revealed, more or less in our backyard within the galaxy, the existence of thousands of so-called exoplanets. As yet there is no evidence for the existence of life, let alone intelligent life, anywhere in the universe except on the earth. At least in our case this too relies critically on clumping (the formation of the earth) and the dissipation of solar energy brought to us by the spangling into space of the sun's high-energy photons. Without them there would be none of the plants on the earth that either directly or through animals provide our food and sustain our life. Critical too is the fact that we and the earth do not get toasted by the solar radiation. Virtually all of it that reaches us is reradiated away into space by low-energy photons, about twenty for every high-energy photon received from the sun. There's a good account of this in Sean Carroll's *From Eternity to Here*, in which he convincingly demonstrates that biological processes on the earth are fully in accord with thermodynamics—"the Second Law is in good shape".

Another example of the emergence of normal thermodynamic behaviour that is fully explicable in normal statistical-mechanical terms once gravitational clumping has created not only stars but also planets and their satellites is the impact of a meteorite on the moon. This creates a crater and a lot of heat which is dissipated in two ways: through radiation into space and by conduction through the surface of the moon.*

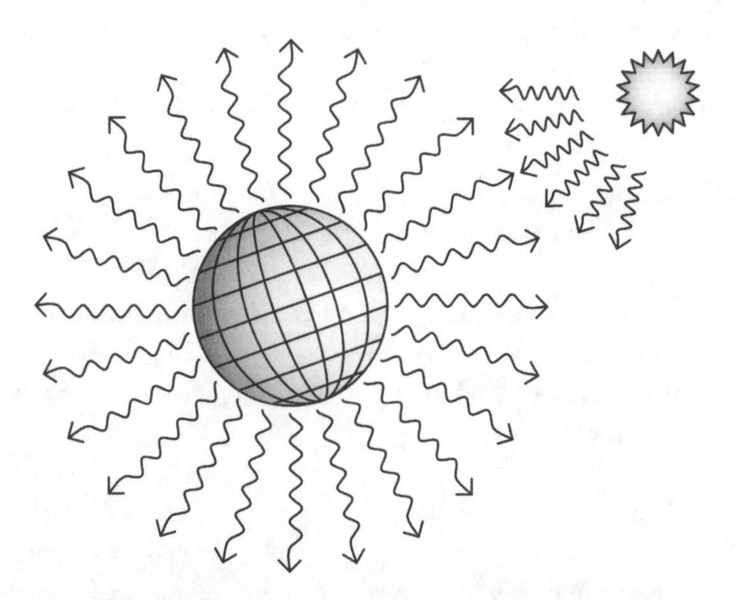

FIGURE 26. Earth receiving high-energy solar photons and emitting low-energy photons.

*More generally, whenever expansion of the universe has allowed subsystems to become isolated they will in general have different temperatures in different regions within them; they will have a tendency to equilibrate. It is not just a meteorite and the moon which can collide; large systems can do that. Indeed, the moon itself is widely thought to have arisen soon after the formation of the solar system from collision of the earth with a Mars-sized planet. This will have led to great creation of heat followed by equilibration. In an expanding universe, equilibration is not a problem; lack of it would be.

However, whether from this one should conclude that in either case disorder is continually increasing is not so clear to me. Fig. 26 is based on Carroll's Fig. 50 and shows one solar photon incoming and twenty terrestrial outgoing. It is beyond gainsaying that the earth, through photosynthesis and reemission of absorbed solar radiation, is spreading (i.e., dissipating) the solar energy, but the earth too, like the sun on a much larger scale, is spangling in a rather nice way. What is more, the information encoded in its outgoing photons can tell distant extraterrestrials rather a lot about us and the rich ecosystem that sustains us. In fact, it is clear that the earth as whole, like conventional radio and TV transmitters, is itself a transmitter which spangles like mad. So are the moon and any hot region created by the impact of a meteorite; even the part of the heat dissipated by conduction through the surface material of the moon, while certainly creating disorder in the familiar entropic sense, does so in an ordered way as the hot spot increases in size. This is an example of microscopic disorder being manifested as macroscopic order. And, of course, impacts of large meteorites have created the many craters we see on the moon. They too are included among the circular or spherical structures and behaviours that become possible in an expanding universe. Galileo's report in *The Starry Messenger* of his observation of them in 1609 with the telescope he had constructed was a worldwide sensation.

I don't think dissipation of mechanical energy will ever cease—Lord Kelvin can lie untroubled next to Isaac Newton in Westminster Abbey—but the radiation emitted by all the stars and black holes in the heavens will never equilibrate. The great range of temperatures, both between the individual emitters and over the course of their lives, is imprinted on the radiation and will not be washed out by expansion of the universe. Heat death in the form of thermal equilibrium is not the end state of the universe. Variety, expressed through shapes and ratios, can increase forever.

Kelvin's dissipation and Clausius's entropy increase are two different ways of saying energy is being spread out. The effect is quite different in confined and unconfined systems. With your thumb, you can press an ink drop on paper into a smudge; the ink, like

Boltzmann's particles in a box, is trapped within the area on which you press your thumb. But nature, the artist, holding a fine pen between fingers and thumb, can draw out the same ink as thinly and as far as inspiration leads, maintaining contrast and with it variety—and creating the finest picture imaginable for eyes that see contrasts. Those contrasts are ratios made manifest. Ratios are the foundation of so much that we experience. The real numbers permit unending extension of ratios—one real number divided by another—all the way to infinity. It can be done with density contrasts on the inside of a sphere. In his nutshell, Hamlet can be king, not of bland infinite space, but of infinite variety.

A filigree is an ornamental work of fine wire (typically gold or silver) formed into delicate tracery. The universe as it now is, with its great chains of star-studded galaxies surrounding vast voids, is a filigree on the most monumental scale. If, in our epoch, the universe were not a filigree that somehow grants conscious experience, W. B. Yeats could not have written his poem 'The Cloths of Heaven':

> Had I the heaven's embroidered cloths,
> Enwrought with golden and silver light,
> The blue and the dim and the dark cloths
> Of night and light and the half-light;
> I would spread the cloths under your feet:
> But I, being poor, have only my dreams;
> I have spread my dreams under your feet;
> Tread softly because you tread on my dreams.

CHAPTER 20

WHAT DOES IT ALL MEAN?

STEVEN WEINBERG ENDS HIS *THE FIRST THREE MINUTES* ON A GLOOMY NOTE. He wrote that particular passage at 30 000 feet flying over Wyoming and looking down on the earth, which "looks very soft and comfortable" but is just "a tiny part of an overwhelmingly hostile universe". He comes to the conclusion that "the more the universe seems comprehensible, the more it also seems pointless". The best comfort he can draw is that "if there is no solace in the fruits of our research, there is at least some consolation in the research itself. . . . The effort to understand the universe is one of the very few things that lifts life a little above the level of farce, and gives it some of the grace of tragedy". Two years after Weinberg published his book in 1977, Freeman Dyson (who many believe should have been awarded the Nobel Prize at the same time as Richard Feynman) published a critical response to Weinberg and declared himself to be, in contrast, a philosophical optimist. In a remarkable, very readable paper, cited in the bibliography and available online, he set out to show that life, intelligence, and civilisation—on an intergalactic scale—could survive forever. In his recent book Brian Greene gives his own, rather less optimistic gloss on Dyson's paper, taking pains to underline what a bleak future the universe does seem to face in the extreme far future. Both authors assume the currently known laws of nature hold into that future and that thinking, as a minimal requirement for meaningful continued existence, involves processing in some localised medium, as certainly happens in our brains.

It seems my parents gave me genes that see the glass, if not so brimming full as for Dyson, at least more than half so. There is no light without the dark. With vision as clear as any, Shakespeare saw all the cruelty of the world and has Hamlet instruct the players to "hold the mirror up to nature". In the most fearful scene in *King Lear*, Gloucester, in his own castle, has his "corky arms" bound fast and a first eye gouged out by Cornwell, who then, despite fatal wounding by an outraged servant, attacks the other eye "lest it see more, prevent it. Out, vile jelly! Where is thy lustre now?" Although the good servant dies and Shakespeare is never Panglossian, surely no one ever described beautiful things better than he. The icicles have lustre too.

In his *A Beautiful Question*, Frank Wilczek asks, "Is the world a work of art?" It is a while since I read the book, but my recollection is that Wilczek does not answer his question with an unqualified affirmative. I also heard him on the BBC a few years ago in which, again as I recall, he said he had not found his way to an understanding of the world and existence that gave him full satisfaction. I wonder if the closest he comes to it is in the belief, expressed in the book, that the scientific revolution began with the discovery of linear perspective in the late Middle Ages; Filippo Brunelleschi is the first artist known to have exploited it in a painting, in 1415. Wilczek chose to make his point not with that painting but with a colour illustration of Perugino's *Delivery of the Keys of Heaven to Saint Peter*, created in 1481–1482. It takes your breath away; it is perhaps the closest thing to seeing heaven. If I failed to persuade you to check online images of the various scientists who have featured in this book, please have a look at Perugino's painting. I bring it as supporting evidence for what I am going to suggest.

As I have worked on this book, a conviction has grown in me: the law which governs the universe is creative and brings things, if not to artistic perfection, at least to precision. It is, however, done subtly and only 'in the fullness of time'. The minimal model's creation of the singleton and Kepler pair is emblematic. Whether one can say there is a goal, that the universe is teleological, is moot; nevertheless, there is a definite tending to precision, though it is never absolutely achieved at any finite time. However, this does not lead to exact symmetry in the end product; instead there is triple realisation of the instruments of metrology. As they tell the time, the particles of the pair are not going round in a perfect circle, but they travel just as effectively in ellipses of the same shape but different sizes; their major axes are rods and compasses that track the singleton's flight. It must, however, be said that although the precision does keep on improving, nothing essentially new happens—the 'needle gets stuck'. This is because Newtonian theory does not allow the number of particles to change. Quantum field theory does allow such change, so that brings in scope for continual creation of novelty. On a vast scale, it seems, that is what the universe does: if my radical proposal for the nature of the big bang is correct, the cosmos begins with near perfect uniformity and evolves to increasingly rich complexity. I think the best argument for that remains the one from Chapter 11: the universe must tend to an attractor, and there is strong evidence from its evolution up to now to suggest that will be one to ever greater complexity.

However, that conjecture will clearly be wrong if the universe ends in a big crunch, which it must if the universe is spatially closed, which I favour for Machian reasons, and has the homogeneous FLRW behaviour for that case. In fact, Dyson had a strong dislike of a closed universe, commenting that it gave him "a feeling of claustrophobia to imagine our whole existence confined to a box" and that he returned "with a feeling of relief to the wide open spaces of the open universe". He thought that it "seems to give enormously greater scope for the activities of life and intelligence". He wanted a universe in which "there will always be new things happening, new information coming in, new worlds to explore, a constantly expanding domain of life, consciousness, and memory". I don't think that is incompatible with a spatially closed universe. A big crunch can be avoided if there is a positive cosmological constant or special matter like a scalar field for which the pressure is negative with magnitude greater than one-third of the energy density (such a scalar field is often invoked as the inflaton that drives inflation). It also seems to me not unreasonable to assume, on grounds of time-reversal symmetry, that the direction of experienced time is determined by eternal expansion of the universe either side of a Janus point. Under that assumption, let us consider what, if anything, of the kind of experience we currently have might also continue forever.

Dyson does not address that question directly; rather, as I indicated, his question was whether the laws of physics as we currently know them will allow processes like those involved when we use our brains to think to continue forever. I am not sure that is the right question to ask; instead, I suspect we have to ask what absolutely minimal condition must be met in the external world if conscious awareness of changing variety is to exist. For something like that to happen, Dyson relies on a localised source of free energy and conjectures that the basis of consciousness is structure rather than the matter that supports it. This opens up the possibility that computers can be conscious, but it does not eliminate the need for whatever is thinking and conscious to shed waste heat. Brian Greene not only likens stars but also brains to steam engines. Continued operation of them along steam-engine lines undoubtedly will get more and more precarious as energy is dissipated ever more thinly throughout the universe.

However, donning now the mantle of a tentative philosophical optimist, I am going to suggest that we can allow some latitude compared with both the restrictions that Dyson and Greene impose on themselves. As regards the laws of nature, I think it is clear that if the universe did have an extraordinarily special birth, as Richard Feynman, Roger Penrose, and many others think is necessary, then the hint we get from Newtonian total collisions and the BKL conjecture suggests that the current laws of nature may be merely the grossest approximation to a delicate interconnectedness of the whole universe. We all have that sense at times. Perhaps nobody expressed it better in words than Shakespeare when he let Bottom, abandoned by the rude mechanicals, awake in the forest and say,

"I have had a most rare vision. I have had a dream—past the wit of man to say what dream it was. Man is but an ass if he go about to expound this dream". Of course, that is what Shakespeare had just done in the play. Bottom continues: "I will get Peter Quince to write a ballad of this dream. It shall be called Bottom's Dream because it hath no bottom". In other words, it has infinite depth. Dyson relies on depth when he compares analog memory and digital computer memory—the latter of necessity has limited capacity, but in principle there is no such restriction on the former. Given the importance I have accorded to dimensionless ratios as the bedrock of physics, I'm pleased to see that Dyson relies entirely on them in his account of how intelligence could survive in a universe that gets ever cooler. Differences are all that Dyson needs. Once scale is taken out of the universe, it is only ratios that remain. Therefore, let us unshackle our imaginations from the current laws of nature. Bottom had a vision; we can have one too.

At the end of Chapter 16, discussing artistic creation, I hazarded the speculation that great performers have a direct awareness of the whole interconnected universe. It would have been better to say that the universe expresses itself through them; they are where consciousness has a location and creates a partial whole within the cosmic edifice. Is there anything better crafted and as complete in itself as a cameo than the icicles, logs, owl, and greasy Joan at the end of *Love's Labour's Lost*? It probably came to Shakespeare in an instant, as if he had just witnessed it. There is also language; it creates a congruence between words and things that don't even exist. In *Henry the Fourth*, Owen Glendower claims he can "call up spirits from the vasty deep", to which Hotspur replies, "Aye, so can I and any man, but will they come when you call them?" In a flash we can at least certainly call up an image of Edward Lear's owl and pussycat who went to sea in a pea-green boat.

Mind and language can change the world; a few hundred words in the New Testament, the ones with most effect, have literally moved mountains, or at least a lot of fine stone from them to build cathedrals. All religious texts have a similar effect. Set against that are the appalling effects of the words of dictators like Hitler.

Where does the power of mind end? I mentioned the theoreticians who seriously contemplate cosmic engineering using black holes to liberate free energy. Earlier I said Carnot's aim of optimising the use of coal could be a distraction from the primary task: finding out what the universe is about. I said that, unless we are gods, what the universe lets us do is secondary. I certainly do not think we are gods, but we are participating actors. One can only wonder what that might mean.

As I said, my second difference from Dyson and Greene is to seek not the ability to process information—that is, to think—but the absolutely minimal condition that must be met in the external world if conscious awareness of changing variety is to exist. That is the most persistent aspect of what is our vivid experience that *time does pass*. In fact, I will hardly go beyond what I already said in *The End of Time* when considering how we can see a kingfisher in flight if, in accordance with the timeless interpretation of quantum gravity considered in that book, the universe is entirely static. Except at the end of Chapter 18, I have said little about quantum issues in this book because it concentrates almost entirely on a pre-quantum description of the universe. However, as regards the problem of consciousness, the only difference I see is that the classical picture gives, as it were, a succession of snapshots that you might take on a walk through the countryside, whereas quantum mechanics forces you to consider, somehow all at once, snapshots taken of all possible views of the landscape. In both cases the snapshots are frozen images. How, with or without motion, the conscious mind becomes cognizant of the material world, whose existence almost all scientists accept, is the 'hard problem'. The best one can say—and it is not a solution—is that there is *psychophysical parallelism*: whatever we experience in consciousness has a counterpart in the physical world. I don't think anybody has gone much beyond what Galileo already said in 1623. I quote from Stillman Drake's translation, available online, of *The Assayer*:

> To excite in us tastes, odors, and sounds I believe that nothing is required in external bodies except shapes, numbers, and slow

> or rapid movements. I think that if ears, tongues, and noses were removed, shapes and numbers and motions would remain, but not odors or tastes or sounds. The latter, I believe, are nothing more than names when separated from living beings, just as tickling and titillation are nothing but names in the absence of such things as noses and armpits.

Galileo was a master of Italian prose and a wit. The tickling and titillation are an example. But that is not why I have included the passage. It's rather to suggest, as I did in *The End of Time* (though for configurations and not yet shapes), that the "slow or rapid movements" should be removed from the external world and taken directly from the instantaneous shape of the universe into consciousness. In movies we know that projection at the correct speed allows our brains, coordinating the information from several successive stills at once, to 'play the movie' for us in consciousness. In his final book, *The River of Consciousness*, the late Oliver Sacks describes some fairly recent neuroscience that I think is concordant with the suggestion I made in *The End of Time*. Reexpressed in terms of shapes, it is that the quantum law of the universe gives high probability to shapes that are very special—namely, the time capsules I talked about in Chapter 1. Each is a single shape with a structure so rich that it encodes the history through which it appears to have come into being. There is little doubt that our brains, through the memories they contain, have a time-capsule nature. Psychologically, memory is one of the most powerful manifestations of time's arrow. It both haunts and gives comfort. Its roles in literature are legion. Who recognised Odysseus when, after twenty years, he finally returned to Ithaca? Not Penelope, but his dog. In brief, my formulation of psychophysical parallelism is that when, for example, we see a kingfisher in flight, the synaptic connections in our brains at any given instant simultaneously encode several 'snapshots' of the kingfisher in different positions and that the collection of them is the counterpart of our conscious experience of seeing motion.

All I would like to add to it now is the comment that the central role of shapes and their differences in the 'plan' of the universe appears to be, through the precision with which the law of the universe acts on us, manifested in the acute awareness of shape, of the layout of things in the world, and of how they can change (at least partly, it seems, through our will) that we humans have and express through our various activities and drives. A few years ago Neil MacGregor, the then Director of the British Museum, created for the BBC a fascinating series, *History of the World in a Hundred Objects*. It took the form of fifteen-minute radio talks, each on one object in the museum chosen to illustrate a particular aspect of human history. Marvelling at not only the skill with which stone axes were made over a million years ago but also at their beauty, MacGregor conjectured that the axes had been created not only for the very necessary business of survival but simultaneously as works of art.

Is there ever going to be a reconciliation of the two cultures, of science and the arts? In a talk given in 1959 in Cambridge, still much discussed, C. P. Snow deplored the two cultures into which the educated world was, in a very one-sided way, divided. He commented on the way people thought to be highly educated "have with considerable gusto been expressing their incredulity at the illiteracy of scientists". This had once or twice provoked him to ask "the company how many of them could describe the Second Law of Thermodynamics. The response was cold: it was also negative". He thought that whereas most scientists had great enthusiasm for and knowledge of the arts, the supposedly educated might even struggle with the concepts of mass and acceleration and have about as much insight into the nature of the world "as their neolithic ancestors would have had". Bearing in mind MacGregor's talks, one might add that perhaps they had rather less. I hasten to say that appreciation of science in the general public has greatly increased, and the snobbery of which Snow complained has correspondingly decreased.

I already commented on the reaction of poets like Blake and Keats, who complained that "all charms fly at the mere touch of cold philosophy". One more poet worth quoting is Yeats, who in 1936 wrote a very readable essay on Bishop Berkeley and his idealistic philosophy, according to which no material world exists at all but only perceptions that God implants in minds. Yeats claimed that "Descartes, Locke, and Newton took away the world and gave us its excrement instead. Berkeley restored the world". This was why the Japanese monk, in his Nirvana Song, could say, "I take pleasure in the sound of the reeds". Yeats concluded: "Berkeley has brought back to us the world that only exists because it shines and sounds. A child, smothering its laughter because the elders are standing round, has opened once more the great box of toys".

If there is to be a reconciliation of the two cultures, I think it must come through mathematics. I incline strongly to the Platonist view of mathematics that Max Tegmark expresses in his *Our Mathematical Universe*. Mathematics, to which one does get used, is in large part nothing but a formal language for speaking about structure, which for its part is almost synonymous with form or shape. Structure is not excrement but what underlies the world, and it is expressed in the way the world shines and sounds. But its critical role as the bones of our back, chest, and limbs is often hidden by the beauty of what is seen. In Botticelli's Venus we see the shape of her naked body, not the bones behind her flesh and the anatomy they support. Let me repeat the much admired part of the quotation from *A Midsummer Night's Dream* that I used to head Chapter 10, on shape space: "As imagination bodies forth the forms of things unknown, the poet's pen turns them to shapes and gives to airy nothing a local habitation and a name". Cézanne could wait for hours until the inspiration came as to where one last spot of green paint should be applied. The whole picture depended on it being the right place. It's just the same with which particular word is put into the structure of a poem and where it is inserted. Poets, artists, composers, theoretical physicists—they are all parts of the universe that evolves from one shape to another. The story of the universe is surely much richer than the evolution of the Newtonian three-body problem, but that already captures the accuracy with which one shape changes into another. Who is to say a supernova explosion expresses the essence of the universe more faithfully than Cézanne's spot of paint? Is the whole universe one common striving?

The question brings us back to being and becoming. Pure shape explains so much and so little. Throughout the book I have ridden the two horses of the extended and reduced representations, sometimes one, sometimes the other. The former is dressed with shape and time as we intuitively think we know them; the latter is reduced to the bare minimum needed to describe mathematically the possible states of the universe. How might Spinoza, *sub specie aeternitatis*, describe that minimum?

The only things that exist are shapes. Classical histories are continuous curves through the space of possible shapes. Both size and duration piggyback on the change of shape. Astronomers see that the shapes of the heavenly bodies were simpler in the past and interpret that as reduction in the size of the universe, down to zero at the big bang. There is no objective warrant for it. You can say that a sphere you hold in your hands has the surface area $4\pi r^2$: the radius r is some fraction of the length of your hand, a ratio. But when you look at a star in the sky you do not see how far it is away from you. You only see angles between it and other stars. The same goes for the celestial sphere; its r^2 is invisible. What you see is its angle measure: 4π. That never changes; it does not get smaller in the past. Dyson's feeling of claustrophobia in a closed universe was misplaced. Our experiential world is always closed—it is the celestial sphere. What gives us the sense of infinite space in that nutshell is not distance but the possibility that the variety within it does not come to an end. Indeed, Dyson himself, when describing the infinite capacity of analog memories, comments that in an expanding universe one can use "a physical quantity such as the angle between two stars in the sky" and that the corresponding memory capacity "is equal to the number of significant binary digits to which the angle can be measured". Please note the reference to an expanding universe, which in terms of actual observations actually means one whose shape changes. In connection with the evidence presented both with and since the account of Kepler pair formation, it seems to me that the history of the universe is one of ever increasing ability to increase the number of significant binary digits that can be measured. The amount of information that can be accessed can grow without limit. In fact, Dyson himself lauded Gödel's proof in 1931 that "the world of pure mathematics is inexhaustible" and commented that if his view of the future is correct it means that "the world of physics and astronomy is also inexhaustible". If, as Tegmark argues, our universe is mathematical, is there any limit to the variety of the shapes it can contain, whether it is open or closed?

To continue with the Spinozan perspective, what the astronomers see is objects at different times in the past. That is meaningful because they belong to shapes that can be ordered sequentially. And in the sweep of cosmic history, the structure of objects and things—the anthropocene earth, gold nuclei, the first stars, the microwave background, and the Harrison-Zel'dovich fluctuations—all get simpler and more (but never perfectly) uniform the further back in time you look. That's matched in the path the evolution curve of the universe takes through shape space into the past: it gets simpler, simplest of all at the big bang. You can see that in the way the splayed histories, traced back to their common origin in Fig. 21, straighten out almost perfectly as they approach the Janus point. Time too is derived: chronology immediately from the succession of shapes, while the difference in complexity of two shapes that the path of the universe passes through in its evolution is a pretty good measure of the duration that clocks measure between them. In our epoch, the time told by the best atomic clocks in the world probably advances very closely in step with the increase of the universe's complexity. Duration is ultimately a measure of the difference of shapes, both on the cosmic scale and within well-isolated emergent subsystems.

That's the picture *sub specie aeternitatis*. The mystery of becoming is twofold. First, what selects the history that is actually realised? In one form or another this problem exists in both classical and quantum universes. Even if many are realised, that still leaves the issue of why they are interesting, and each follows a path for which no reason can be found in the structure of shape space. This is the problem of an effect without a cause—the violation of Leibniz's principle of a sufficient reason. Of course, we would not be here if it were not violated. Most mysterious of all is consciousness.

Even if, as seems eminently reasonable to me, our brain does contain in any instant several 'stills' of the kingfisher in flight, where is the projector that turns them into motion in our minds?

Back to both cosmology and art one last time and how much effect tiny differences can have. I have a strong recollection of reading many years ago about a special feature in the construction of the Parthenon, the temple dedicated at the time of Pericles in the fifth century BCE to Athena on the Acropolis of Athens. To avoid a jarring effect, the uppermost platform was not made absolutely level but, as was discovered only in modern times, was constructed with a slight upward curve as part of a circle with its centre high in the sky. On checking this out I found online at MichaelTFassbender.com an only slightly different explanation: were the uppermost platform, which is 228 feet long, exactly level, it "would seem to dip in the middle, so it was built to ascend instead, by just over four inches . . . creating the impression that it was perfectly level . . . The Parthenon was built to seem perfect to the eye, rather than being perfect in fact". My pocket calculator tells me the dip of four inches, as a ratio of 228 feet, is one part in 786, or not much more than one part in 1000. In whatever way the Harrison-Zel'dovich fluctuations may have arisen, they do seem to be the origin of all the structure in the universe. The magnitude of the fluctuations to which they gave rise in the cosmic microwave background is one part in 100 000; on astronomical scales that is not all that different from the relative dip in the Parthenon's platform. The Parthenon itself has been one of the greatest sources of inspiration to artists and thinkers for centuries. Besides purloining surreptitiously a word or two from Shakespeare, I might have done better than toying with Alpha or Beta as the name for the location of the Einsteinian big bang. An acropolis is the summit of a city. Perhaps I should have called it The Acropolis.

This book, as so many do, took on a life of its own as I wrote it. Besides its purely scientific aspiration to understand the origin of structure in the universe, it seems to have become, at least in part, a song of thanks to the cosmos and the fact that I, like you, am a participant in whatever it does. I think there are many more things besides the effort to understand the universe that lift life a little above the level of farce. We can do and enjoy so many things. The universe is indeed made of stories, some of which we ourselves create even if tragedy lurks around the corner. In fact, adding to the speculation about the origin of artistic creation with which I concluded Chapter 16, could it be that the best guide to the true nature of the universe is to be sought in art? Should we look to Shakespeare rather than Newton and Einstein?

A theme running through the book has been the variety that can be observed everywhere in the universe. Three or four years ago I was asked to give the eulogy at the funeral of a carpenter who lived in the same village as I do. John was a true country-lover, born and bred in a village not far away. His especial delight was in the patterns one could see in nature, especially in birds. He would spend hours in the fields with his binoculars. I therefore thought nothing would fit better than to include the poem 'Pied Beauty' by Gerald Manley Hopkins:

Glory be to God for dappled things—
For skies of couple-colour as a brinded cow;
For rose-moles all in stipple upon trout that swim;
Fresh-firecoal chestnut-falls; finches' wings;
Landscape plotted and pieced—fold, fallow, and plough;
And áll trádes, their gear and tackle and trim.

All things counter, original, spare, strange;
Whatever is fickle, freckled (who knows how?)
With swift, slow; sweet, sour; adazzle, dim;
He fathers-forth whose beauty is past change:
Praise him.

It was only after I sat down that I realised the poem was even more apt than I had said in introducing it. The celebrant mentioned something I knew well but had forgotten: John was colour-blind, but he always said he never had the feeling his lack of colour vision in any way diminished his immense delight in nature. There was so much patterned variety everywhere.

LIST OF ILLUSTRATIONS

NOTES

THE NOTES FOR THE various chapters of the book that now follow serve two purposes: first, to amplify in somewhat more technical detail the main substantive claims made in the book, and second, to provide references for readers who wish to consult books and original research papers related to the various topics covered in this book. I suspect most readers will only be interested in the books on time discussed in the notes for Chapter 1. In the more technical notes I generally presume prior knowledge such as what a tensor is; this is especially true in the notes for Chapter 8, which not only anticipate the background to some of the later chapters but also, as it were, put the dynamical backbone into the book. The bibliography that follows the notes is arranged in alphabetical order of the author (first in the case of multiple authors except in the case of the papers by Chen and Carroll and by Gibbons and Ellis, in which the conventional order was presumably reversed because the first author contributed the major part) and chronologically in the case of more than one entry for the same author. After that there is the index, in which I have included not only technical terms but also sometimes words or brief phrases used once that you may recall. An example is 'owl', which prompted reference in a footnote to a book you may want to consult (I recommend it). In line with common practice, the notes are not indexed.

Chapter 1: Time and Its Arrows

There are numerous books for the general reader that concentrate on the problem of time's arrows and the second law of thermodynamics, but they all treat the topic in broadly the same way and do not make the central distinction that I do between confined and unconfined systems. Together with brief comments, I list here in chronological order of their publication the authors of only the books with which I am familiar and which more or less directly relate to the subject of this book. Full details of the books are given in the bibliography. Sean Carroll's book contains an ample and excellent bibliography of books both on time in general as well as on its experienced direction.

Hans Reichenbach's 1956 book is mainly intended for philosophers of science.

Paul Davies is a rare example of a theoretician who has combined valuable research with a string of solid books for the lay reader. His first book on time (1974) was completed just before the discovery of black-hole entropy.

Roger Penrose is a very clear writer and covers a wealth of topics unrelated to time. I have included all of his four books for lay readers (1989, 2005, 2010, 2016) in the bibliography because the issue of entropy and time's arrows appears prominently in them all, especially the controversial *Cycles of Time*. As I say in Chapter 3, Penrose gives a beautiful discussion of equilibration in a confined system with a phase space of bounded measure but, as I argue in Chapter 5, uses it in the context of the universe, for which I think the assumption is questionable. However, for a very different reason to Penrose's I do, in Chapters 16, 17, and 18, propose what I think could be a realisation of the very special initial condition for the universe that Penrose advocates.

Peter Coveney and Roger Highfield's 1990 book covers a good range of topics and is specifically intended for the lay reader.

Huw Price's 1996 book is for both the general reader and the philosopher of science.

Paul Davies's second book (1995) contains this splendid self-deprecatory comment in the preface: "You may well be even more confused about time after reading this book than you were before. That's alright; I was more confused myself after writing it". The comment emphasises how confusing both ordinary people and theoretical physicists find the subject of time.

David Albert's 2000 book contains a discussion of the past hypothesis.

Sean Carroll's 2010 book is very readable and covers well numerous aspects of the modern theory of time omitted entirely in my book. His is the book where you should look for information about inflation and the multiverse.

Lee Smolin, who has had considerable success with his books for the general reader, argues that time is fundamental and contrasts his view with mine; if for no other reason, I recommend his 2013 book. In fact, in support of his case Lee uses shape dynamics and the distinguished notion of simultaneity (the CMC foliation, as in Fig. 9) that my collaborators and I have developed. In shape dynamics (see Chapter 8, especially the final part), a primary aim is to derive dynamics without a primary notion of time, which is argued to be only emergent.

Like me, Carlo Rovelli argues that time is relational, but he does not contemplate the possibility, illustrated in Fig. 9 and discussed in the third part of Chapter 8, that the dynamical structure of general relativity might define a notion of universal simultaneity in the universe. His discussion in his 2019 book of the revolution in the nature of time associated with Einstein's theory of relativity (both special and general) is a good account, but his notions of thermal and perspectival time, which deal respectively with the problems that arise in the attempts to quantise gravity and explain our experienced arrow of time, are controversial.

Brian Greene's 2020 book is the most recent one that deals with the arrow of time. He comes to a very different conclusion than I do about the ultimate destiny of the universe.

As regards the minute violation of time-reversal symmetry mentioned at the start of Chapter 1 and the explanation of why there was not complete mutual annihilation of matter and antimatter in the early universe (referred to at the end of the chapter), the most frequently cited paper is Andrei Sakharov's 1979 work.

The famous Russian physicist became a Soviet dissident, having earlier been a key person in the development of the USSR's hydrogen bomb.

Chapter 2: Bare-Bones Thermodynamics

For either the general reader or professional scientists there are surprisingly few books on the history of thermodynamics. I'm inclined to recommend the book by Carnot that laid its foundations; make sure you get the one with additional comments by the editor E. Mendoza. For students who want to see what came out of Carnot's work as developed by Thomson and Clausius, nothing beats Fermi's slim booklet. In 1867, Clausius published a book containing all nine of his papers on phenomenological thermodynamics, and it was very soon translated into English; in 1879, he published a textbook with the same title. This can be bought, having been made available by the Internet Archive. Lemons's book, which includes extensive quotations from Carnot's book and the papers of Thomson and Clausius, is especially good on the definition of the second law. However, it seems to me Lemons is wrong on page 42 of his book in his claim that Carnot relied on the impossibility of perpetual motion to show that no heat engine operating irreversibly could outperform his reversible one. Carnot's discussion was whether one *agent* or another would perform better in a reversible engine. This is clear from the passage in Carnot's book that Lemons quotes on page 53: "Is the motive power of heat invariable in quantity, or does it vary with the agent employed?" It was the adaptation of Carnot's argument without the notion that caloric exists that led both Clausius and Thomson independently to their formulations of the second law (Thomson's is frequently referred to as 'Kelvin's formulation'). Lemons cites approvingly the book by Cardwell; by all accounts it is excellent, though at the time of writing I have not had a chance to look at it. For physics students Jacobson's article is brief and full of fascinating material very clearly presented; the volume of essays in honour of Jacob Bekenstein in which it appears will give you an idea of most modern approaches to entropy in cosmology. Clifford Truesdell's book is for specialists; I only looked at parts of it and got the impression he was failing to see the forest for the trees. There is a wonderful amount of information about Thomson (Lord Kelvin) and his age in the collection of essays about him edited by Flood, McCartney, and Whitaker. The Einstein quotation at the end of the chapter is from his 'Autobiographical notes'. The Eddington quotation is from his 1927 Gifford Lectures.

Chapter 3: Statistical Mechanics in a Nutshell

Thanks to the efforts of Stephen Brush, the history of statistical mechanics is very well covered. Two of his books are especially recommended. The title of the one published in 1976 repeats the title that Clausius gave to the first of his papers on the kinetic theory of gases. The collection of classic papers with commentaries published in 2003 is invaluable. It has one omission that is a pity: Boltzmann's second paper of 1877, in which he spelled out the count-of-microstates idea. The statistical-mechanics companion to Fermi's *Thermodynamics* is Erwin Schrödinger's *Statistical Thermodynamics*.

Chapter 4: Boltzmann's Tussle with Zermelo

All the papers mentioned in this chapter, above all the two each of Zermelo and Boltzmann (along with Thomson's 1874 paper), are available in Brush's translations in his book published in 2003. Brush also translated Boltzmann's *Lectures on Gas Theory*, which, in sec. 90 ("Application to the Universe"), seems to contain his final thoughts on the origin of time asymmetry.

Chapter 5: The Curious History of Boltzmann Brains

All the papers and books mentioned in this chapter can be found under the respective authors in the bibliography. I am grateful to Patrick Dürr for drawing my attention to the papers by Eddington in 1931 and von Weizsäcker in 1939, in which I found the reference to the paper by Bronstein and Landau. Along with Eddington's blunt warning to anyone who challenges the second law of thermodynamics that I quote in the discussion of Clausius's famous claim that the entropy of the universe tends to a maximum, his statement "Shuffling is the only thing Nature cannot undo" is also to be found in his 1927 Gifford Lectures.

Chapter 6: The Janus Point

Although never to my knowledge used to discuss the arrow of time, the Newtonian *N*-body problem, being in many cases a very good approximation to Einstein's general theory of relativity, plays a very important role in modern cosmology. For example, virtually all calculations of the growth, from very small initial inhomogeneities, of large-scale structure in the universe employ Newtonian theory on the background of an expanding universe. At the conceptual level, two very interesting papers are those of Gibbons and Ellis, which draw attention to how many features of cosmology traditionally studied using Einstein's theory can be illuminated and even better described using Newton's theory. Prior to their papers not many books or papers used Newtonian dynamics to model the cosmos. An exception is Hermann Bondi's *Cosmology* of 1952, republished in 2010 with an introduction by Ian Roxburgh. It gives the best argument available in the 1950s for the Copernican (or cosmological) principle, both in the weaker form that at any given epoch the universe should, on a suitable averaging scale, look the same everywhere in space and in the form that the founders of the steady-state theory of the universe called the perfect cosmological principle, which says that the universe should also present the same appearance at all times. This idea very soon became untenable. It was through reading Bondi's book in 1963 that I learned about Ernst Mach's critique of Newton's notion of absolute space just at the time I had independently come to the conviction that time is abstracted from change. Both issues play decisive roles in Chapter 8.

The proof that the square of the root-mean-square length has the behaviour depicted on the left in Fig. 1 (the behaviour depicted on the right is discussed later in the book) is based on what N-body specialists call the Lagrange-Jacobi relation but is more widely known as the virial theorem. It was in fact Clausius who first proved the theorem, realised its significance, and introduced its name; his paper is included in Brush's anthology of classic papers listed in the bibliography. Proofs of the relation can be found in Landau and Lifshitz (1960), Barbour, Koslowski, and Mercati (2014), and Gibbons and Ellis (2013). The condition required for the relation to hold for some dynamical system is that the potential V, assumed to depend on the coordinates x_i, is homogeneous of some degree k, which means that $V(ax_i) = a^k V(x_i)$, where $a > 0$ is some constant and k can be any number. The virial theorem relates the potential V, the centre-of-mass energy E_{cm}, and the second time derivative of the centre-of-mass moment of inertia I_{cm}, which apart from an inessential constant factor is an alternative expression for the square of the root-mean-square length l_{rms} and is half the trace of the moment-of-inertia tensor. The explicit form of the relation is $\ddot{I}_{\text{cm}} = 4E_{\text{cm}} - 2(2 + k)V$. For the Newton potential $k = -1$, we get $\ddot{I}_{\text{cm}} = 4E_{\text{cm}} - 2V$. It is now important that V_{New} is always negative, so if E_{cm} is either zero or positive $\dot{I}_{\text{cm}} > 0$, which means that $\dot{I}_{\text{cm}} > 0$ is a monotonically increasing function that varies from $-\infty$ to $+\infty$ for both directions of Newtonian time, which in turn means that the plot of I_{cm} and with it the square of ℓ_{rms} is a curve that is U-shaped concave upwards, as in Fig. 2. Half of $\dot{I}_{\text{cm}}$ is equal to

$$D = \sum_{i=1}^{i=N} \mathbf{r}_i^{\text{cm}} \cdot \mathbf{p}_i^{\text{cm}},$$

where $\mathbf{r}_i^{\text{cm}}$ and $\mathbf{p}_i^{\text{cm}}$ are the centre-of-mass coordinates of the particles. The quantity D has the dimensions of angular momentum but is conserved only if the potential is homogeneous of degree $k = -2$. So far as I know, it had not been given a name in dynamics until I coined the name *dilatational momentum* for it in Barbour 2003.

In N-body literature it is generally denoted J, probably from the initial letter of Jacobi.

I have not made an exhaustive search through the literature, but to the best of my knowledge the only clear anticipation, of which my collaborators and I were unaware in 2014, of the Janus-point idea in the form we proposed is due to Jennifer Chen and Sean Carroll in their 2004 paper I mention in Chapter 5. Their unambiguous aim, precisely explained with U-shaped entropy curves, was to solve the problem of time's arrow and they did so using an idea based on the then (and still) popular notion of the multiverse. There is a relatively brief discussion of the idea in Carroll's book *From Here to Infinity*, published in 2010. I won't go into the details, which, being based on the multiverse, are at least somewhat speculative, since the multiverse is too. I will, however, mention that very soon after the 2004 paper Carroll discussed in various lectures a much simpler model in which a finite number of particles move purely inertially in space. In this case too the size of the system of particles, measured by the root-mean-square length ℓ_{rms}, passes through a unique minimum and grows to infinity in both directions of Newtonian time (the 'potential' for inertial motion is homogeneous of degree zero and the Lagrange-Jacobi relation holds for it, leading to a U-shaped curve). This is a Janus-point model but without gravity. It therefore shares with the gravitational model its most distinctive property but lacks the complexity-increasing effects of gravity that play a dominant role in this book. The model featured in an article in *The New Scientist* of 13 January 2016 but has not been published in a scientific journal. A much earlier partial anticipation of a 'Janus-point' cosmological model, with the timeline of the universe divided into two halves at a distinguished time $t = 0$, was proposed by Andrei Sakharov (mentioned in the notes to Chapter 1) and described in Sakharov's 1980 paper. So far as I can understand the brief paper, there is no attempt to have proper continuation of evolution through $t = 0$. Sakharov simply notes that the second law of thermodynamics will be time-reversal symmetric if the entropy S of the universe increases in both directions of time from $t = 0$. Sakharov's model was a partial stimulus to the very different Janus cosmological model that Jean-Pierre Petit has been developing for many years and describes in an English-language video on YouTube. See also the mirror-type universe of Boyle, Finn, and Turok (2018) inspired by modern particle physics.

Chapter 7: The Minimal Model

Dziobek (1892) proved that the three-body problem with vanishing angular momentum is planar; a proof can also be found in Sundman (1913); the standard monograph on the three-body problem is Marchal (1990). Since ℓ_{rms} must tend to infinity under the condition of vanishing energy assumed, along with vanishing angular momentum, for the minimal model, at least one of the particles must escape to infinity. This means that either a singleton must separate from the other two particles, which then form a Kepler pair, or all three particles must escape to infinity. This is called *parabolic escape*, by analogy with what happens in the two-body problem, for which the orbits are elliptic for negative energy, parabolic for zero energy, and hyperbolic for positive energy. It is rather remarkable, and not trivial to prove, that parabolic escape can only happen if the triangle which the particles form as they tend to infinity is equilateral; see Sundman (1913), Chazy (1918), Marchal (1990), and the 1998 paper of Chenciner, who advises that the statement that Tauberian estimates are not needed is incorrect. This is an exceptional, so-called *zero-measure* case; it is an example of the remarkable *central configurations*, which play a decisive role in Chapter 16.

Chapter 8: The Foundations of Dynamics

The notes for this chapter are more extensive because of the critical role that Machian relational ideas play throughout the book; the notes anticipate and are intimately related to the notion of shape space and shape dynamics, which make their first appearance in Chapter 10. You may wish to return to these notes after you have read some of the later chapters. In these notes, I also want to highlight the contributions of the various researchers who have either directly or indirectly contributed to the overall picture presented in the book. The notes also justify, to the extent that I think is possible, the assumption, quite contrary to the standard interpretation of general relativity, that the very structure of Einstein's theory suggests that a distinguished notion of simultaneity does play a role in the theory as in the schematic CMC foliation of spacetime on the right in Fig. 9.

For anyone who wants all the proper mathematical details, the reference is my collaborator Flavio Mercati's 2018 book *Shape Dynamics*; it contains an extensive bibliography. A book with a somewhat different perspective but much overlap is *The Problem of Time* by Edward Anderson (2017), who along with Brendan Foster, Bryan Kelleher, and Niall Ó Murchadha played a critical role in the initial development of shape dynamics from the barest of ideas, including a depiction of the three-body shape space in box 3 of my *The End of Time*, published in 1999. In the first part of these notes I outline the development of Machian ideas before the issue of scale became a central issue. In the context of the three-body problem, that means that the scale variable is treated as a physical variable with the same status as the shape variables. After that, I sketch the development of shape dynamics, which began in 1999.

To begin now with the period before that, the issue of whether motion is relative or absolute has long been a subject of great interest to philosophers of science. Leibniz's objections to Newton's notions of absolute space and time can be found in *The Leibniz-Clarke Correspondence*, Alexander (1956). Mach's critique is in his *The Science of Mechanics*, originally published in 1883; Mach (1960) is the English translation of the 9th German edition of 1912. Also worth consulting are his comments at the end of Mach (1911), which is the translation (available online) of a booklet first published in 1871. The proceedings of the Mach's principle conference held in Tübingen in 1993, edited by Barbour and Pfister (1995), contain a wealth of information about the various interpretations and attempts to implement Mach's principle in the twentieth century. There were strikingly many independent rediscoveries of attempts, made both before and after Einstein's creation of general relativity, to implement the principle directly in a least-action formalism involving only the separations between the particles, the first being due to the Austrian engineer Wenzel Hoffmann in 1904. The form that was then repeatedly rediscovered for a system of N particles in Euclidean space was based on the Lagrangian

$$\sum_{i<j} m_i m_j \dot{r}_{ij}^2 / r_{ij}$$

and first proposed by Reissner in 1914 and then by Schrödinger in 1925, not long before his creation of quantum wave mechanics. Schrödinger was well aware that the theory would lead to an anisotropic effective inertial mass that would result in perturbations of the planetary orbits; using the then estimated mass of the galaxy and our distance from its centre, he concluded that only a slight increase in the accuracy of the relevant data would rule out such a theory. In fact, the vastly more sensitive Hughes-Drever atomic-physics experiment described in Will's paper in the Tübingen proceedings completely rules out such a theory. Of this I was unaware when my own independent rediscovery of the Hoffmann-Reissner theory was published in Barbour (1974). This led to my collaboration with Bruno Bertotti, who was an expert on the experimental aspects of gravity research and acutely aware of the mass-anisotropy problem; we eventually developed the notion of best matching, which does not lead to the mass-anisotropy problem. Aspects of best matching are discussed in Gryb (2010) and Gomes (2011).

The main novelty of my 1974 paper, retained in the 1982 paper with Bertotti, was the elimination of time as an independent variable and the recovery of time (duration) as an abstraction from change. It was only through the intervention of Karel Kuchař in 1980 that it transpired I had rediscovered Jacobi's principle of 1837. Jacobi had introduced it purely formally to make the variational principle introduced by Euler and Lagrange mathematically consistent. Lanczos (1970) contains a very good account of Jacobi's principle, which is traditionally used to transform Newtonian dynamics for one fixed value of the energy into a geodesic principle for the orbit of the system in its configuration space, after which the speed in orbit is found by using the energy conservation law. The action in Jacobi's principle is expressed using an arbitrary time label and is *reparametrisation invariant*. As a result, the momenta in the theory satisfy a *quadratic constraint*; it lies behind the theory of duration that can be deduced from Jacobi's principle and leads to the notion of vertical stacking in Fig. 8, as outlined in Barbour (2009) together with material that goes beyond the account in Barbour and Bertotti (1982), which also introduces the notion of horizontal stacking. The outcome of best matching is called the intrinsic derivative in the 1982 paper; I introduced the terminology 'best matching' in my two 1994 papers listed in the bibliography. I describe the part played by Mach's ideas in Einstein's creation of general relativity in Barbour (1990). My conclusion is that Mach's critique of Newton's concepts of absolute space and time provided the stimulus to Einstein to keep persevering in the creation of his theory but that the example of Maxwell's theory of electromagnetism, the need to ensure local energy-momentum conservation, and above all the preexisting mathematical formalism of Riemannian geometry based on the indefinite metric that Hermann Minkowski had introduced in his famous lecture of 1908 were what actually led him to his theory.

Einstein's coining of 'Mach's principle' is in Einstein (1918) (with an extract in Barbour and Pfister [1995]). His formulation requires that the geometry of spacetime, which he calls the *G*-field, should be "*completely* determined by the masses of the bodies", i.e., by the energy-momentum tensor $T_{\mu\nu}$ of the matter. This is mathematically inconsistent because the definition of $T_{\mu\nu}$ itself involves the geometry. His statement that "it will be immediately recognised" that Mach's principle cannot be right is in Einstein (1949) and cited on page 186 of Barbour and Pfister (1995). This, in Einstein's view, was because Mach had formulated his idea in the context of the Newtonian theory of masses that interact instantaneously at a distance and that dynamics in this form had been completely superseded by Maxwell's and his own field theories. In fact, an appropriately formulated Mach's principle can, through best matching, be applied equally well to particle dynamics and field theory, as outlined in Chapter 8 and spelled out fully in Mercati (2018); I will say more about that in a moment.

First, I want to mention the critical importance, discussed by me in Barbour (2010), of an analysis by Poincaré in chapter VII of his *Science and Hypothesis* (1905) of the failure of predictability that arises from Newton's use of absolute space. Suppose that in the N-body problem with known mass ratios of the particles the separations r_{ij} of the particles together with the rates of change of those separations are known at some instant. Poincaré says that if motion really were purely relative, that information should be sufficient to predict the future evolution of the separations. But such data contain no information about the angular momentum in the system, the effect of which then shows up in the subsequent evolution. Because this manifestly happens in the solar system, Poincaré said he felt obliged to accept Newton's argument for the existence of absolute space even though, as a philosopher, he found this violation of the relativity of motion "repugnant". For some reason, despite widespread discussion of Mach's ideas, Poincaré did not consider the possibility that his requirement might be satisfied, not for the solar system but for the universe.

Although shape space is not introduced in the body of the book until Chapter 10, it will be helpful to introduce it here because of its intimate connection with relational Machian ideas. One arrives at the notion of shape space, both in particle dynamics and in field theory, in several steps. The starting point in particle dynamics is the standard Newtonian configuration space, which first appeared in this book in Chapter 3 in the discussion of Boltzmann's introduction of what we now call the count of microstates. In the Newtonian configuration space particle positions are generally defined by means of Cartesian coordinates in an inertial frame of reference. In an N-body system it has $3N$ coordinates. The *relative configuration space* (RCS), defined in Barbour (1974), is the $(3N - 6)$-dimensional reduced configuration space obtained by quotienting with respect to Euclidean translations and rotations. It is important that it is not a subspace of the configuration space but a distinct space. To arrive at shape space, which has $3N - 7$ dimensions, one further step needs to be taken, which is to quotient by dilatations. If one uses the N-body problem as a model universe and insists there can be no external rod or clock to make measurements within it, one can ask what kind of dynamics might be realised in it.

To consider what possibilities exist, it is very interesting to see what form general Newtonian solutions take in shape space. As a preliminary for that consideration, I first need to mention the *velocity decomposition theorem*, which is proved in Saari (2005) and states that at any instant the centre-of-mass total kinetic energy can be decomposed into three orthogonal parts that respectively represent the amounts of kinetic energy in overall rotation, overall expansion, and pure change of shape. From them two dimensionless ratios can be formed, neither of which can be encoded in a shape and a direction in shape space, which I will call shape Cauchy data. It is also not possible to encode the direction of the angular momentum vector relative to the instantaneous N-body configuration, so there is a further shortfall by two of Newtonian data that cannot be encoded in shape Cauchy data. One last such datum arises from the value of the energy. If it is not zero, the ratio of the kinetic to the potential energy cannot be encoded in a point and direction in shape space. The upshot is this: The shape Cauchy data for general Newtonian solutions consist of a point and direction in shape space together with *five* extra data. The one associated with the energy arises because an external time is employed, and the three associated with angular momentum exist because orientation is absolute. In a relational model of the universe, these four can be set to zero initially and will remain zero because time does not occur explicitly in the action principle, and the Newton potential only depends on the interparticle separations and not the orientation. But if one wishes to retain the Newton potential one cannot dispense with the last dimensionless ratio, which is the ratio of kinetic energy in overall expansion to the amount in change of shape. This is the ratio that determines the bending parameter (or creation measure).

Shape dynamics in the form in which it currently exists arose from an initial attempt to describe the universe without an additional parameter of this kind. I found such a theory in 1999; it describes the interaction of *N* point particles and was eventually published as Barbour (2003). I think its main interest now is that it shows what is wrong with such a theory. However, it may have value in highlighting the fact that in an expanding universe which has vanishing angular momentum there exists a quantity that has the same dimensions as angular momentum and is a measure of the overall expansion of the universe; so far as I know, it was first given a name, *dilatational momentum*, in my paper of 2003 (as pointed out in the notes to Chapter 5, *N*-body specialists simply tend to denote it by *J*). If one takes the Machian relational approach seriously, it is hard not to wonder whether there might be a deep connection between the microscopic and macroscopic aspects of the universe, reflected in the fact that the parameter which above all characterises the microscopic world is Planck's quantum of action, which also has the dimensions of angular momentum. Since the development of the theory of primordial nucleosynthesis (discussed later in the book and brilliantly described by Stephen Weinberg 1993 in his 1993 book *The First Three Minutes*) it has been recognised that what happens at the microscopic level when elementary particles interact with each other has an extremely important effect on the overall behaviour of the universe. However, a direct connection between the dilatational momentum of the universe and Planck's constant would establish a much more profound interconnection.

Returning to the development of shape dynamics, my paper of 2003 in fact created a theory in which the dilatational momentum is always exactly zero and the (nominal) size of the universe remains constant forever. That this was a defect was not originally recognised, and the idea of creating an analogous theory of evolving three-dimensional geometry was developed by myself and Niall Ó Murchadha in 1999 in the preprint cited in the bibliography. Its key concepts can be said to come directly from Riemann's paper of 1854 that generalised Euclidean geometry to curved geometries that are Euclidean only in infinitesimal regions. The analogue of the Newtonian configuration space before quotienting with respect to translations and rotations is Riem, the space of all possible Riemannian three-metrics (with signature +++) on a closed three-manifold. When this is quotiented by diffeomorphisms, all metrics that differ merely by the coordinates used to describe them are identified; they describe one and the same *three-geometry*, the complete set of which is *superspace* (it has no connection with supersymmetry in particle dynamics). One further quotienting by three-dimensional conformal transformations, which leaves the angles between intersecting curves unchanged but changes the local volume element, leads to *conformal superspace*. Later, in Chapter 17, I call the group whose transformations consist of three-dimensional diffeomorphisms and conformal transformations the *geometrical group*. It is the generalisation of the similarity group to Riemannian geometry. The 1999 paper of Ó Murchadha and myself defines a geodesic theory on conformal superspace. Somewhat developed, it appeared as Anderson et al. (2003). Even more than with the example of my particle theory, published as the companion paper Barbour (2003), I now see its main value as highlighting what is wrong with such a theory: it does not allow the universe to expand or interesting structure to form and evolve.

The defect is rectified by what seems to be a tiny change. Instead of seeking a theory whose action is invariant under arbitrary three-dimensional conformal transformations, one allows only those transformations that keep a nominal volume of three-dimensional space constant. Before I explain the significance of this restriction to volume-preserving conformal transformations, I need to say something about a major advance in relativity theory that had actually occurred, with no Machian motivation in mind, three decades earlier through the work of James York and Niall, who was at that time a PhD student with York; in this connection their two most important papers are York (1972) and Ó Murchadha and York (1973). They found an extremely effective way to solve what is known as the initial-value problem of general relativity. Einstein's theory is based on ten tensor field equations $G_{\mu\nu} = \kappa T_{\mu\nu}$, where κ is a constant that depends on the units employed and μ and

ν take the values 0, 1, 2, 3 (0 for time, 1, 2, 3 for the three spatial components); at each spacetime point the quantities $G_{\mu\nu}$ determine the state of the spacetime geometry, while the quantities $T_{\mu\nu}$ determine the state of the matter. There are only ten equations since the tensors $G_{\mu\nu}$ and $T_{\mu\nu}$ are symmetric. The field equations form what is called an elliptic-hyperbolic system because the 00, 01, 02, 03 equations do not involve the time at all and one must find data that contain only spatial derivatives and satisfy these equations; only then can one use the remaining hyperbolic equations to evolve the initial data. Such a situation already exists in Maxwell's theory in which data that satisfy one such equation, the Gauss constraint, must be found before one can begin to use the equations that determine the evolution of the electromagnetic field.

This is already a nontrivial problem but it is much harder in general relativity. In the Hamiltonian dynamical formalism of general relativity developed by Arnowitt, Deser, and Misner (1962), the dynamical variables are, at each space point on a spacelike hypersurface, a 3×3 symmetric tensor g_{ij}, $i,j = 1, 2, 3$ that describes the intrinsic three-geometry and its conjugate momentum, which is also described by a 3×3 tensor p^{ij}. The 01, 02, 03 Einstein equations are linear, analogous to the Gauss constraint in Maxwell's theory, and take the form $p^{ij}{}_{;j} = 0$ (the momenta are *transverse*). From a Machian point of view they arise from best matching with respect to three-dimensional diffeomorphisms and are analogous to the condition of vanishing angular momentum *at each space point*. It is the (algebraic) 00 equation that is really important and difficult; it is quadratic in the momenta and, at each space point, analogous to the single quadratic constraint that arises in Jacobi's principle. Its existence leads to the notorious 'problem of time' in quantum gravity as described in my *The End of Time* and Anderson (2017). The great difficulty in finding data that satisfy all four constraints arises from the fact that they are coupled and the 00 equation is quadratic.

Building on earlier work by Lichnerowicz (1944), York first showed that one could find momenta that are not only transverse but also *traceless*, which means they satisfy the further condition

$g_{ij}p^{ij} = 0$. He then showed that such data are invariant with respect to conformal transformations, under which one can make an arbitrary position-dependent change of the local volume factor (the determinant of g_{ij}) and a simultaneous transformation of the momenta p^{ij} in such a way that they remain transverse and traceless with respect to the transformed g_{ij}. One has merely a 'skeleton' three-geometry that defines angles between curves that intersect on the underlying three-manifold but without any notion of volume. The transverse traceless data, to which York could add a spatially constant trace, allow one to specify the volume element completely freely at each space point. The beautiful thing then is that one can 'put flesh on the skeleton' because the 00 equation has now been decoupled from the three linear constraints and can be used to determine a unique value of the local volume factor. This means that one has found initial data that satisfy all four constraints. The form in which the 00 constraint has been cast is called the Lichnerowicz-York equation; for it there always exists a unique solution (Ó Murchadha and York [1973]) for appropriately specified boundary conditions or without boundary conditions in the spatially closed case. Some form of York's method is the basis of virtually all numerical work in general relativity; it played a critical role in calculations of the gravitational waves resulting from the merger of black holes.

A few more comments are needed to tie the preceding summary to the Machian interpretation of general relativity outlined in the final part of Chapter 8. First, the condition that the momenta should be both transverse and traceless is exactly what is to be expected as a result of best matching with respect to both diffeomorphisms and conformal transformations. Next, the initial data as found by York's method can be evolved to build up spacetime in an arbitrary foliation, as on the left in Fig. 12. However, the initial data themselves correspond to data on a surface of constant mean extrinsic curvature (CMC foliation), as on the right of Fig. 9. York showed that evolution of the initial data once found can be continued (at least in an open neighbourhood) in a CMC foliation in the framework of the Hamiltonian formalism of Arnowitt, Deser, and Misner by solving a *lapse-fixing equation* for which a unique solution always exists. This has very interesting implications for the volume. It turns out that there are several different ways in which one can solve the Lichnerowicz-York equation in the spatially closed case; they are described in Ó Murchadha (2005). In one, an initial total volume of space is specified; the Lichnerowicz-York equation then determines how that total volume is distributed at each local space point. If one uses the CMC lapse-fixing equation to evolve the initial data, the consequence is that all the local volume elements at each space point increase or decrease at the same fractional rate everywhere across space. The rate of this monotonic change is called *York time*. The dilatational momentum can be used as an analogous time variable in the particle model.

A further positive feature of York's conformal method is that one can couple the bare conformal geometry to all currently known forms of matter and still solve the equations of the corresponding initial value problem (see Isenberg, Ó Murchadha, and York [1976] and Isenberg and Nester [1977]). This shows that there is a perfectly Machian determination of the local inertial structure of spacetime that involves the matter *and* gravitational degrees of freedom everywhere in space; this puts a first-principles Machian interpretation on Isenberg (1995) and Isenberg and Wheeler (1980), in which it is simply noted, without the best-matching and Jacobi-type motivation, that the York method yields such a determination of the local inertial structure in an Einsteinian spacetime.

A further comment relates to a certain inconsistency in York's method as originally formulated. He added the spatially constant trace to the transverse momenta with everywhere vanishing trace because then the resulting Lichnerwicz-York equation could be solved uniquely. His seemingly inconsistent procedure, found by trial and error, could be justified by the fact that the resulting mathematical equation had good properties. A more satisfactory procedure finally appeared in Anderson et al. (2005); expressed in Machian terms, it replaced best matching with respect to full conformal transformations by ones that keep the total volume of space constant. The constancy at the level of best matching unfreezes what would be constancy of the volume under the dynamical evolution. The result is a completely consistent formulation of a modified Lichnerowicz-York equation which respects the fact that the volume of the universe is not constant in general relativity.* The difference between my geodesic particle model of 2003 and general relativity is striking. The particle model has only a single scale variable and it cannot change. In the relational *N*-body model it can. In general relativity, there is a single evolving global scale variable (or rather ratio of a scale variable in the form of York time) and infinitely many local scale variables, which all evolve.

In Chapter 8 I said that when all the inessentials are stripped from the shape-dynamical core of general relativity we will see it naked like Michelangelo's David with "all its inner strength revealed". Conformal three-geometries brought as near as is possible to congruence by best matching constitute the shape-dynamical core; the inessentials are the two things that are supposed to be the very essence of Einstein's theory: space and time glued together as spacetime. But a conformal geometry contains no notion of distance; it is the solution of the Lichnerowicz-York equation that puts the position-dependent local scale factor into the conformal geometry as 'padding' and creates something one can call space. As for time, in the form of local proper time (Minkowski's notion of duration), the solution of the lapse-fixing equation puts it as a 'prop' at each space point of the successive conformal three-geometries and yields a notion of elapsed time. Mach's claim that "time is an abstraction at which we arrive by means of the changes of things" is vindicated with the most perfect precision.

Finally, to what extent does this discussion suggest that the mathematical structure of general relativity leads to a notion of simultaneity defined throughout a spatially closed universe? In the absence of a proper theory of quantum gravity, I do not think this is readily answered. The problem resides in the enigmatic nature of the 00 Einstein field equation—the quadratic Hamiltonian constraint of

---

*I should say that, through my fault, the variational technique employed in Anderson et al. (2005) is invalid, as was pointed out to me by Karel Kuchař; the condition that is claimed to hold throughout the solution in fact does so only at the limits of integration. Fortunately, the key change in York's method to volume-preserving (rather than full) conformal transformations remains valid.

the formalism developed by Arnowitt, Deser, and Misner (1962). Its structure is what allows Einstein's field equations to take exactly the same form however one chooses to foliate spacetime (as indicated symbolically on the left in Fig. 9). The question is this: is there any truly sufficient reason to break the foliation invariance? For me the most persuasive argument is due to Dirac in his two papers of 1958 and 1959 included in the bibliography. In the first he developed, independently of Arnowitt, Deser, and Misner, the Hamiltonian formalism of general relativity and considered the implications of the foliation invariance associated with the form of the Hamiltonian constraint.* Dirac was hugely impressed by the form general relativity takes when cast as a theory of dynamically evolving three-geometry when cast in the simplest possible form. He noted that "Hamiltonian methods, if expressed in their simplest form, *force one to abandon the four-dimensional symmetry*". He said he was "inclined to believe from this that four-dimensional symmetry is not a fundamental property of the physical world". In his later paper of 1959, Dirac argued that foliation invariance must lead, at the least, to a very awkward quantum theory of gravity since it would enforce unacceptable inclusion of gauge degrees of freedom. In the Hamiltonian formulation, the theory has four such gauge degrees of freedom per space point; three can be eliminated by means of the linear constraints $p^{ij}{}_{;j} = 0$. The question is then how the fourth gauge freedom could be eliminated. Dirac argued that the only possibility which used the fundamental dynamical variables of the theory was to impose the condition of vanishing trace: $g_{ij}p^{ij} = 0$. This is called the maximal slicing condition. In a spatially closed universe it can only hold at an instant at which the volume of the universe passes through a maximum. Had Dirac wanted to develop a quantum theory for a closed universe, he would have had

---

* In connection with this mention of the Hamiltonian constraint, I would like to mention, somewhat tangentially, the paper 'Relativity without relativity' (Barbour, Foster, and Ó Murchadha [2002]), which derives (without presupposing special relativity) first vacuum general relativity, next a universal light cone that is enforced when matter is coupled to dynamical three-geometry, and finally gauge theory if 1-forms are coupled. This is almost entirely due to the special structure of the Hamiltonian constraint.

to choose the condition York employed to solve the initial-value problem and require a spatially constant trace. In just a few pages, Dirac all but formulated the main results of shape dynamics as my collaborators and I understand them.

Two more things need to be said. York's method can be used to generate a great number of solutions of Einstein's equations, but there is no guarantee that they can be continued in CMC foliation beyond an initial open neighbourhood. Moreover, one and the same spacetime might well admit two or more intersecting CMC foliations. There is, however, one interesting situation that may be relevant in this discussion, which is that it may be possible to specify York-type initial data in the immediate vicinity of the big bang. In such a case, all the relevant information that determines the spacetime would be encoded in CMC-type data even if such a foliation could not be continued indefinitely. This possibility is discussed in Chapter 17 and the notes to it.

Chapter 9: The Complexity of Shapes

Koslowski, Mercati, and I introduced complexity and the Janus-point idea in Barbour, Koslowski, and Mercati (2014), from which Fig. 12 has been extracted. We also cast our ideas in terms of shape dynamics.

Chapter 10: Shape Space

All of the implications of the introduction of shape space and the argument that the history of the universe should be interpreted as a succession of shapes are presented in the body of the book and in the extended notes to Chapter 8. The neutral measure mentioned in the chapter is constructed from the metric based on the Newtonian kinetic metric for displacements subject to the constraint of vanishing angular momentum and divided by the centre-of-mass moment of inertia (equivalently the square of the root-mean-square length). This is 'neutral' because it does not involve any force-generating potential, and it defines 'inertial motion' of the universe in its 'shape space'.

Chapter 11: A Role Reversal and Chapter 12: This Most Beautiful System

Lagrange's demonstration that Newtonian dynamics can be re-expressed in an illuminating form in the three-body problem is extended to the *N*-body problem in Albouy and Chenciner (1998).

Chapter 13: Two Kinds of Dynamical Laws

Noether's paper, two translations of it, and Dirac's book (1964) are cited in the bibliography.

Chapter 14: A Blindfolded Creator

The paper by Gibbons, Hawking, and Stewart and the response to it by Schiffrin and Wald are both in the bibliography.

As noted in the Notes to Chapter 10, the neutral measure on shape space is defined by means of its mass-weighted natural metric, which is the best-matched Newtonian kinetic metric made dimensionless through division by the moment of inertia. The majority of shape space, according to this measure, is occupied by Poissonian distributions of points. This can be seen by observing that pre-shape space, the submanifold in R^{3N} of configurations with unit moment of inertia and appropriate mass rescaling, is a unit $3N - 1$ sphere with the round metric on it. Then the actual shape sphere can be obtained by quotienting by SO(3) rotations, which are the ones that rotate the three coordinates of each particle into each other the same way for all particles. But, as far as the metric is concerned, this quotient does not change the measure because SO(3) rotations are isometries of the natural metric. This means that the projection on actual shape space of rotationally invariant regions of pre-shape space have a volume that is identical to the volume that the natural metric attributes to them in pre-shape space.

All of this means that 'to measure space' it is merely necessary to measure regions in the unit $3N - 1$ sphere. Now, from Marsaglia's article (1972) on the n-sphere (and specifically the section on generating uniformly distributed random points on the sphere), the fastest way is to generate random points in R^n with a Gaussian distribution centred at the origin and then normalise the coordinate vectors so that they lie on the unit sphere. But this, for large n, generates a Poisson distribution, because each coordinate of each particle is generated independently with a random number extraction completely uncorrelated from the others. The Gaussian profile doesn't change the fact that for a large number N of particles the correlations between particle coordinates are those of a Poisson process and to a significant degree uniform, the predominance of Poissonian distributions increasing strongly with N. An important thing to remark in connection with this note, for help on which I thank Flavio Mercati, is that Fig. 11, in which the distribution on the left is Poissonian, shows that other distributions exist which are significantly more uniform. Their existence will turn out to be critical later in the book.

Chapter 15: An Entropy-Like Concept for the Universe

The notion of entaxy and the suggestion that an entropy-like quantity for the universe should decrease rather than increase is in Barbour, Koslowski, and Mercati (2015).

Chapter 16: A Glimpse of the Big Bang?

Relevant papers that can be consulted on total collisions are in the first place Sundman (1913), Chazy (1918), and Chenciner (1998). Two authoritative textbooks are Marchal (1990) and Wintner (1941). There are beautiful introductions to central configurations in Gibbons and Ellis (2013), Saari (2011), and both of Moeckel's (2014) papers. The planar central configurations of Fig. 17 are taken from mimeographed notes of Moeckel with the title 'Some relative equilibria of N equal masses, $N = 4, 5, 6, 7$' that do not appear to have been published and which I somehow acquired a few years ago. The striking results of numerical simulations to find low complexity central configurations are in the 2003 paper of Battye, Gibbons, and Sutcliffe; Fig. 18 (repeated as Fig. 25) was found numerically by Flavio Mercati and is similar to the central configuration of 500 equal-mass particles in their paper, which explains in a very illuminating way how Newton's famous potential theorem is the origin of the beautiful spherical shapes of central configurations with near maximal value of the shape potential. That there are only discrete central configurations in the four-body problem was proved using computer software by Hampton and Moeckel (2006) and analytically by Albouy and Kaloshin (see their 2012 paper for further results for the five-body problem and references to their four-body study).

Chapter 17: The Big Bang in General Relativity

Much of the discussion of this chapter is based on Koslowski, Mercati, and Sloan (2018), from which Fig. 23 (slightly modified) is taken. The original formulation of the BKL conjecture is in Belinskii, Khalatnikov, and Lifshitz (1970); an excellent source of information about it and the evidence that supports it can be found in Ashtekar, Henderson, and Sloan (2011). The conjecture is proved for analytical solutions of Einstein's equations in Andersson and Rendall (2001). Specifically, they prove "the existence of a family of spacetimes depending on the same number of free functions as the general solution which have the asymptotics suggested by the Belinskii-Khalatnikov-Lifshitz proposal near their singularities"; this is what I mean by the 'goodly number' mentioned in the main text. In connection with my comments about CMC foliations in the notes to Chapter 8, Andersson and Rendall comment in the conclusion of their paper that "a neighbourhood of the singularity can be covered by a foliation consisting of constant mean curvature hypersurfaces". Thus, all the information in spacetimes allowed by the kind of law of the universe that I am proposing is encoded in such a foliation obtained by best matching.

There is a very nice review of the Yamabe theorem and the many conclusions that can be drawn from it in Ó Murchadha (1989). It may also be mentioned here that spatial metrics divide into positive, zero, and negative classes. The method for solving the initial-value problem in general relativity (discussed in the notes for Chapter 8) that Lichnerowicz developed in 1944 works only for the positive class. As developed by York, the method works for all three classes on a spatially closed manifold. There is a comment about this in Ó Murchadha (2005).

Chapter 18: The Large-Scale Structure of the Universe

Alan Guth's excellent account of the development of the theory of cosmological inflation is in Guth (1998). Hollands and Wald (2002) propose an alternative to inflation together with a measured and very readable account of the strengths and weaknesses of the theory of inflation. For a recent review, see Vázquez, Padilla, and Matos (2020). The discussion of the appearance of Harrison-Zel'dovich fluctuations in real space draws on Gabrielli, Joyce, and Labini (2002) and Gabrielli et al. (2003); the illustration in Fig. 24 is essentially fig. 1 of the second paper. I am very grateful to Michael Joyce for drawing my attention to these two papers.

Chapter 19: The Creation of Structure in the Universe

The quotation used to illustrate the role of free energy in the creation of structure in the first part of this chapter is taken from Dyson, Kleban, and Susskind (2002). My account of the standard explanation of complexity growth despite the assumed growth of entropy in the universe owes much to Frautschi's 1982 paper; it presents the assumed principles clearly and is very good on the estimation of entropy increase through emission of quanta, above all photons, through the radiation of various bodies in the universe. More recently, Lineweaver (2013) gives a broadly similar account but without Frautschi's details of individual entropy-increasing processes. Although Gold (1962) is widely credited, correctly I believe, with being the first person to attribute the arrow of time to expansion of the universe, his paper came before the great flowering of cosmology that was ushered in by the discovery of the cosmological microwave background. He attributes the arrow of time almost exclusively to radiation of stars into space made possible by expansion of the universe and its resolution of Olbers's paradox of the dark sky at night. This is what I have called spangling; Gold takes no account of the formation of structure through gravitational clumping. While he discusses the issue of advanced and retarded radiation without making the point that I do about the dilution of the CMB background, he does, like me, point out that in a deterministic theory the conditions that prevail at a given instant will be evolved forward in time.

Although, as is evident from Chapter 16, I do take seriously the possibility that the universe began in a very special state that could be argued to have an exceptionally low entropy as advocated by Feynman and Penrose, it remains my belief that the evidence for any such special origin is largely hidden by early effective equilibration and the resulting observed arrow of time is due to the expansion of the universe and the structure formation it makes possible.

BIBLIOGRAPHY

Albert D (2000), *Time and Chance*, Harvard University Press.

Albouy A and Chenciner A (1998), 'Le problème des *N* corps et les distances mutuelles', *Invent. Math.* **131**, 151.

Albouy A and Kaloshin V (2012), 'Finiteness of central configurations of five bodies in the plane', *Ann. Math.* **176**, 535.

Albrecht A and Sorbo L (2004), 'Can the universe afford inflation?' *Phys. Rev. D* **70**, 063528; arXiv:hep-th/0405270v2.

Alexander H (1956), *The Leibniz-Clarke Correspondence*, Manchester University Press.

Anderson E (2017), *The Problem of Time*, Springer.

Anderson E, Barbour J, Foster B, and Ó Murchadha N (2003), 'Scale-invariant gravity: geometrodynamics', *Class. Quant. Grav.* **20**, 1571; arXiv:0211022/gr-qc.

Anderson E, Barbour J, Foster B, Kelleher B, and Ó Murchadha N (2005), 'The physical gravitational degrees of freedom', *Class. Quant. Grav.* **22**, 1795; arXiv:0407104/gr-qc.

Andersson L and Rendall D (2001), 'Quiescent cosmological singularities', *Comm. Math. Phys.* **218**, 479; arXiv:0001047/gr-qc.

Arnowitt R, Deser S, and Misner C (1962), 'The dynamics of general relativity', arXiv:gr-qc/0405109 (consolidation of several papers published in the 1950s).

Ashtekar A, Henderson A, and Sloan D (2011), 'Hamiltonian formulation of the Belinskii-Khalatnikov-Lifshitz conjecture', *Phys. Rev.* D **83**, 084024; arXiv:1102.3474v2/gr-qc.

Barbour J (1974), 'Relative-distance Machian theories', *Nature* **249**, 328.

Barbour J (1990), 'The part played by Mach's principle in the genesis of relativistic cosmology', in: *Modern Cosmology in Retrospect*, eds. Bertotti B, Balbinot R, Bergia S, and Messina A, Cambridge University Press.

Barbour J (1994), 'The timelessness of quantum gravity: I. The evidence from the classical theory', *Class. Quant. Grav.* **11**, 2853.

Barbour J (1994), 'The timelessness of quantum gravity: II. The appearance of dynamics in static configurations', *Class. Quant. Grav.* **11**, 2875.

Barbour J (1999), *The End of Time*, Oxford University Press.

Barbour J (2003), 'Scale-invariant gravity: particle dynamics', *Class. Quant. Grav.* **20**, 1543.

Barbour J (2009), 'The nature of time' (essay in FQXi competition), arXiv:0903.3489.

Barbour J (2010), 'The definition of Mach's principle', *Found. Phys.* **40**, 1263; arXiv:1007.3368.

Barbour J (2012), 'Shape dynamics. An introduction', in: *Quantum Field Theory and Gravity*, eds. Finster F et al., Springer; arXiv:1105.0183/gr-qc.

Barbour J and Bertotti B (1982), 'Mach's principle and the structure of dynamical theories', *Proc. Roy. Soc. Lond. A* **382**, 285 (available as PDF at platonia.com).

Barbour J, Foster B, and Ó Murchadha N (2002), 'Relativity without relativity', *Class. Quant. Grav.* **19**, 3217; arXiv:gr-qc/0012089.

Barbour J, Koslowski T, and Mercati F (2014), 'Identification of a gravitational arrow of time', *Phys. Rev. Lett.* **113**, 181101; arXiv:1409:0917/gr-qc.

Barbour J, Koslowski T, and Mercati F (2015), 'Entropy and the typicality of universes', arXiv:1507.06498v2/gr-qc.

Barbour J and Ó Murchadha N (1999), 'Classical and quantum gravity on conformal superspace', arXiv:9911071/gr-qc.

Barbour J and Ó Murchadha N (2010), 'Conformal superspace: the configuration space of general relativity', arXiv:1009.3559/gr-qc.

Barbour J and Pfister H (1995), *Mach's Principle: From Newton's Bucket to Quantum Gravity*, Einstein Studies, Vol. 6, Birkhäuser.

Battye R, Gibbons G, and Sutcliffe M (2003), 'Central configurations in three dimensions', *Proc. Roy. Soc. A* **459**; arXiv:0201101/hep-th.

Belinskii V, Khalatnikov I, and Lifshitz E (1970), 'Oscillatory approach to a singular point in relativistic cosmology', *Adv. Phys.* **19**, 525.

Boltzmann L (1995), *Lectures on Gas Theory*, Dover (German originally published in 1898; this translation by Stephen Brush originally published by the University of California Press in 1964).

Bondi H (1952), *Cosmology* (reprinted with introduction by Ian Roxburgh by Dover in 2010).

Boyle L, Finn K, and Turok N (2018), 'CPT-symmetric universe', *Phys. Rev. Lett.* **121**, 251301; arXiv: 1803.08928 [hep-th] (see also arXiv: 1803.08930 [hep-th]).

Bronstein M and Landau L (1933), 'On the second law of thermodynamics and the universe', in: *Collected Papers of L. D. Landau*, ed. Ter Haar D (1965), Pergamon.

Brush S (1976), *The Kind of Motion We Call Heat: A History of the Kinetic Theory of Gases in the 19th Century*, Elsevier.

Brush S (2003), *The Kinetic Theory of Gases; An Anthology of Classic Papers with Historical Commentary*, Imperial College Press.

Cardwell D (1971), *From Watt to Clausius: The Rise of Thermodynamics in the Early Industrial Age*, Cornell University Press.

Carnot S (1824), *Reflections on the Motive Power of Fire*, Dover.

Carroll S (2010), *From Eternity to Here*, Penguin.

Chazy J (1918), 'Sur certaines trajectoires du problème des *n* corps', *Bull. Astron.* **35**, 321.

Chenciner A (1998), 'Collisions totales, mouvements complètement paraboliques et réduction des homothéties dans le problème des *n* corps', *Regul. Chaotic Dyn.* **3**, 93.

Clausius R (1867), *The Mechanical Theory of Heat* (translations of papers), John Van Voorst.

Clausius R (1879), *The Mechanical Theory of Heat*, trans. Browne W R, Macmillan.

Clifford, W (1879), *Lectures and Essays*, ed. Stephen L and Pollock L, Macmillan.

Coveney P and Highfield R (1990), *The Arrow of Time*, W H Allen.

Davies P (1974), *The Physics of Time Asymmetry*, Surrey University Press.

Davies P (1995), *About Time*, Penguin.

Dirac P (1958), 'The theory of gravitation in Hamiltonian form', *Proc. R. Soc. Lond.* A **246**, 333.

Dirac P (1959), 'Fixation of coordinates in the Hamiltonian theory of gravitation', *Phys. Rev.* **114**, 924.

Dirac P (1964), *Lectures on Quantum Mechanics*, Belper Graduate School of Science (reprinted by Dover in 2001).

Dyson L, Kleban M, and Susskind L (2002), 'Disturbing implications of a cosmological constant', *Journal of High Energy Physics* **2002**; arXiv: hep-th/0204212.

Dziobek O (1892), *Mathematical Theories of Planetary Motions*, Register.

Dziobek O (1900), 'Über einen merkwürdigen Fall des Vielkörperproblems', *Astron. Nachr.* **152**, 3.

Eddington A (1927), Gifford Lectures, published in 1928 as *The Nature of the Physical World*, Cambridge Scholars.

Eddington A (1931), 'The end of the world: From the standpoint of mathematical physics', supplement to *Nature*, no. 3203.

Einstein A (1917), 'Cosmological considerations on the general theory of relativity', translated from the German in *The Principle of Relativity* with other papers by Perrett W and Jeffrey G B with notes by Sommerfeld A, Dover (1952).

Einstein A (1918), 'Prinzipielles zur allgemeinen Relativitätstheorie', *Ann. Phys.-Berlin* **55**, 241.

Einstein A (1949), 'Autobiographical notes', in: Schilpp P A, ed., *Albert Einstein: Philosopher-Scientist*, Library of Living Philosophers.

Fermi E (1935), *Thermodynamics*, Dover.

Feynman R (1965), *The Character of Physical Law*, MIT Press (republished with foreword by Frank Wilczek in 2017).

Flood R, McCartney M, and Whitaker A, eds. (2008), *Kelvin: Life, Labours and Legacy*, Oxford University Press.

Frautschi S (1982), 'Entropy in an expanding universe', *Science* **217**, 593.

Gabrielli A, Jancovici B, et al. (2003), 'Generation of primordial cosmological perturbations from statistical mechanical models', *Phys. Rev. D* **67**, 043506; arXiv:0210033v2/astro-ph.

Gabrielli A, Joyce M, and Labini F (2002), 'Glass-like universe: Real-space correlation properties of standard cosmological models', *Phys. Rev. D* **65**, 083523; arXiv:0110451v2/astro-ph.

Gibbons G and Ellis G (2013), 'Discrete Newtonian cosmology', *Class. Quant. Grav.* **31**, 025003; arXiv:1308.1852v2 [astro-ph].

Gibbons G and Ellis G (2015), 'Discrete Newtonian cosmology: perturbations', *Class. Quant. Grav.* **32**, 055001; arXiv:1409.0395v1 [gr-qc].

Gibbons G, Hawking S, and Stewart J (1987), 'A natural measure on the set of all universes', *Nucl. Phys. B* **281**, 736.

Gold T (1962), 'The arrow of time', *Am. J. Phys.* **30**,403.

Gomes H (2011), 'Classical gauge theory in Riem', *J. Math. Phys.* **52**, 082501.

Greene B (2020), *Until the End of Time*, Penguin.

Gryb S (2010), 'A definition of background independence', *Class. Quant. Gravity* **27**, 21508; arXiv:1003.1993/gr-qc.

Guth A (1998), *The Inflationary Universe: The Quest for a New Theory of Cosmic Origins*, Vintage.

Hampton M and Moeckel R (2006), 'Finiteness of relative equilibria of the four-body problem', *Invent. Math.* **163**, 289.

Hollands S and Wald R (2002), 'Essay: an alternative to inflation', *Gen. Rel. Grav.* **34**, 2043; arXiv:021001/hep-th (see also arXiv:0205058v2/gr-qc).

Isenberg J (1995), 'Wheeler-Einstein-Mach spacetimes', in: Barbour and Pfister (1995).

Isenberg J and Nester J (1977), 'Extension of the York field decomposition to general gravitationally coupled fields', *Ann. Phys.-New York* **108**, 368.

Isenberg J, Ó Murchadha, and York J (1976), 'Initial-value problem of general relativity. III. Coupled fields and the scalar-tensor theory', *Phys. Rev. D* **13**, 1532.

Isenberg J and Wheeler J (1980), 'Inertia here is fixed by mass-energy there in every W model universe', in: *Relativity, Quanta, and Cosmology*, Pantaleo M and DeFinis F, eds., Johnson Reprint.

Jacobson T (2019), 'Entropy from Carnot to Bekenstein', in: *Jacob Bekenstein: The Conservative Revolutionary*, World Scientific; arXiv:1810.07839/physics.hist-ph.

Koslowski T, Mercati F, and Sloan D (2018), 'Through the big bang: continuing Einstein's equations beyond a cosmological singularity', *Phys. Lett. B* **778**, 339; arXiv:1607.02460.

Lanczos C (1970), *The Variational Principles of Mechanics*, Dover.

Landau L and Lifshitz E (1960), *Mechanics*, Pergamon.

Lemons D (2019), *Thermodynamic Weirdness: From Fahrenheit to Clausius*, MIT Press.

Lichnerowicz A (1944), 'L'intégration des équations de la gravitation relativiste et le problème des n corps', *J. Maths. Pures Appl.* **23**, 37.

Lineweaver C H, Davies P C W, and Ruse M, eds. (2013), *Complexity and the Arrow of Time*, Cambridge University Press.

Mach E (1911), *History and Root of the Principle of Conservation of Energy*, Open Court (available online: [pdf]jhu.edu).

Mach E (1960), *The Science of Mechanics*, Open Court.

Marchal C (1990), *The Three-Body Problem*, Elsevier.

Marchal C and Saari D (1976), 'On the final state of the *n* body problem', *J. Differ. Equations* **20**, 150.

Marsaglia G (1972), 'Choosing a point from the surface of a sphere', *Annals of Mathematical Statistics* **43**, 645, available at projecteuclid.org/euclid.aoms/1177692644.

McGehee R (1974), 'Triple collisions in the collinear three-body problem', *Invent. Math.* **27**, 191.

Mercati F (2018), *Shape Dynamics: Relativity and Relationalism*, Oxford University Press.

Moeckel R (2014), 'Central configurations', *Scholarpedia*, **9**(4), 10667.

Moeckel R (2014), 'Lectures on central configurations', available at www-users.math.umn.edu/rmoeckel/notes/CentralConfigurations.pdf.

Noether E (1918), 'Invariante Variationsprobleme', *Kgl. Ges. Wiss. Nachr. Göttingen, Math. Phys. Kl.* **2**, 235; translated in Tavel M (1971), "Application of Noether's theorem to the transport equation", *Transport Theory and Statistical Physics* **1**, 271–285, and Kosmann-Schwarzbach E (2011), *The Noether Theorems*, Springer. See also Brading K and Castellani E, eds. (2003), *Symmetries and Noether's Theorems*, Cambridge University Press.

Norton J (2015), "You are not a Boltzmann brain", http://philsci-archive.pitt.edu/17689.

Ó Murchadha N (1989), 'The Yamabe theorem and general relativity', in: *Conference on Mathematical Relativity*, Bartnik R, ed. Proceedings of the Centre for Mathematical Analysis Vol. 19; available at https://projecteuclid.org/euclid.pcma/1416335848.

Ó Murchadha N (2005), 'Readings of the Lichnerowicz-York equation', *Acta Phys. Polon. B* **36**; arXiv:0502055/gr-qc.

Ó Murchadha N and York J (1973), 'Existence and uniqueness of solutions of the Hamiltonian constraint of general relativity', *J. Math. Phys.* **14**, 1551.

Penrose R (1989), *The Emperor's New Mind*, Oxford University Press.

Penrose R (2005), *The Road to Reality*, Jonathan Cape.

Penrose R (2010), *Cycles of Time*, Bodley Head.

Penrose R (2016), *Fashion, Faith and Fantasy*, Princeton University Press.

Poincaré H (1905), *Science and Hypothesis*, Walter Scott Publishing (translation of *Science et hypothèse* [1902]).

Reichenbach H (1956), *The Direction of Time*, University of California Press (reprinted by Dover, 1999).

Rovelli C (2019), *The Order of Time*, Riverhead.

Saari D (2005), *Collisions, Rings, and Other Newtonian N-Body Problems*, American Mathematical Society.

Saari D (2011), 'Central configurations—a problem for the 21st century', [pdf]psu.edu.

Sakharov A (1979), 'The baryonic asymmetry of the universe', *Sov. Phys. JETP* **49**, 594.

Sakharov A (1980), 'Cosmological models of the universe with reversal of time's arrow', *Sov. Phys. JETP* **52**, 349.

Schiffrin J and Wald R (2012), 'Measure and probability in cosmology', *Phys. Rev. D* **86**; arXiv:1202.1818/gr-qc.

Schrödinger E (1946), *Statistical Thermodynamics* (1952 edition republished in 1989 by Dover).

Schrödinger E (1960), *Space-Time Structure*, Cambridge University Press.

Smolin L (2013), *Time Reborn: From the Crisis in Physics to the Future of the Universe*, Houghton Mifflin Harcourt.

Sundman K (1913), 'Mémoire sur le problème des trois corps', *Acta Math.* **36**, 105.

Truesdell C (1980), *The Tragicomical History of Thermodynamics, 1822–1854*, Springer.

Vázquez J, Padilla L, and Matos T (2020), 'Inflationary cosmology: from theory to observations', arXiv:1810.09934v2 /astro-ph.CO.

von Weizsäcker C (1939), English translation in 'The second law and the difference between the past and the future', in: *The Unity of Nature*, Farrar Straus Giroux (1980).

Weinberg S (1993), *The First Three Minutes*, Basic Books (updated edition of book first published in 1977).

Wilczek, F (2008), *The Lightness of Being*, Basic Books.

Wilczek, F (2015), *A Beautiful Question*, Penguin.

Wintner A (1941), *The Analytical Foundations of Celestial Mechanics*, Princeton University Press.

York J (1972), 'Role of conformal three-geometry in the dynamics of gravitation', *Phys. Rev. Lett.* **28**, 1082.

INDEX

JULIAN BARBOUR is the author of the highly regarded *The Discovery of Dynamics* and the bestseller *The End of Time*. He received his PhD in physics from the University of Cologne in 1968. He is a past visiting professor of physics at the University of Oxford and lives on the edge of the scenic Cotswolds, UK.